The Human Impact of
Climate Uncertainty

The Human Impact of Climate Uncertainty

Weather Information, Economic Planning, and Business Management

W.J. Maunder

ROUTLEDGE
London and New York

First published 1989
by Routledge
11 New Fetter Lane, London EC4P 4EE

© 1989 W.J. Maunder
Printed and bound in Great Britain by
Billings & Sons Limited, Worcester

British Library Cataloguing in Publication Data

Maunder, W.J. (William John), *1932–*
    The human impact of climate uncertainty: weather information,
    economic planning and business management
    1. Climate. Changes. Socioeconomic aspects
    I. Title
    304.2'5

ISBN 0-415-04076-0

Library of Congress Cataloging in Publication Data

Maunder, W.J.
        The human impact of climate uncertainty: weather
    information, economic planning, and business management /
    W.J. Maunder.
            p.   cm.
        Bibliography: p
        Includes index.
        ISBN 0-415-04076-0. — ISBN 0-415-04077-9 (pbk.)
        Climatology — Social aspects. 2. Weather — Social aspects.
    3. New Zealand — Climate. I. Title.
    QC981.M445 1989
    304.2'5–dc19   89-3461
                                                                                              CIP

*To Melva
in appreciation of her
continued help and support*

CONTENTS

IX. WEATHER RELATIONSHIPS

FIGURES

TABLES

FOREWORD

There has been an astonishing increase, particularly over the last few years, in public attention regarding the costs of climate change. This has largely stemmed from the consolidation of our knowledge concerning the implications of ozone depletion and of increases in greenhouse gases in the atmosphere. As I write, hardly a day goes by without a feature or news item on the greenhouse effect in a national United Kingdom newspaper; and it is natural that such interest should have rubbed off on the cost of shorter-term climatic variations - the droughts, floods, cold spells, etc. that have always been a part of climate, have always taxed our lives, and for which we shall continue to pay a price.

We should allow Dr Maunder a wry smile at the public's and policy-maker's new-found interest in the impact of climate and weather, because more than any other scientist, he has championed this cause for more the past 20 years. Maunder's first book, *The Value of the Weather* (1970), pioneered a new field - economic climatology. It emphasized the costs and benefits that are to be realized if we respond appropriately. While that work was addressed to scientists, in the next book he was speaking more to the business community. *The Uncertainty Business* (1986) argued that the atmosphere is an 'elite resource' which deserves careful husbandry and the consultation between business, policy-making, and climatology at the highest levels of national planning.

In this book, Maunder takes his message to the public showing, in a wealth of examples that range from transportation and electricity generation to dairy farming and wool production, that careful study of the links between weather and production and consumption is needed to provide the basis for informed, successful decision-making.

In every field of science there is the cosy coterie of specialists which frowns on the popularizer of their field. But the potential effects of greenhouse-gas induced climate change and the costs of climatic hazards *now* are too great for this to deserve any credit. The issue should be one that captures the attention of the scientist, the politician, and the businesss community alike. Maunder is making an important contribution in capturing that wider field.

Dr Martin L. Parry
Atmospheric Impacts Research Group
Department of Geography
University of Birmingham, Birmingham
England

PREFACE

If you wait until the wind and the weather is just right, you will never sow anything and never harvest anything.   *Ecclesiastes 11:4*

From the many comments I have had since my first book *The Value of the Weather* was published in 1970, it is evident that many of the 'ideas' expressed in that book were ahead of their time. However, during the last 10-15 years there has been an increasing amount of activity in what could be called the 'non-scientific' aspects of meteorology/climatology. These include the growing recognition both by the more influential newspapers and journals, and by the more influential politicians, of the real role of weather and climate in economic, social, strategic, and political activities.

When *The Value of the Weather* was written, it was among other things a reaction against the considerable emphasis on 'meteorological processes' which were more than adequately covered in many books. The book was also written to emphasize the political, planning, legal, sociological, and overall 'value' of the weather, and in particular the value of *information* about the weather. Nearly 20 years on the meteorological processes are very adequately covered in various books, but the 'non-scientific' aspects are still neglected by many meteorologists and climatologists, as well as by many economists, agriculturalists etc. Accordingly the economic, social, political, planning, and legal aspects still need to be very much emphasized - even though many advances have been made, particularly by the UN agencies, some national meteorological services, and many meteorological consultants.

Since *The Value of the Weather* was published in 1970, the acceptance of the atmosphere as a variable and 'elite' resource that may be despoiled, modified, monitored, ignored, used, and forecast has become a reality. Nevertheless, this realization exists only in certain nations, communities, institutions, and companies. Indeed, there remain many people - both within and external to the field of weather and climate - who still consider that the weather and climate-resource is devoid of value, and therefore has no place in operational decision-making. My second book *The Uncertainty Business: Risks and Opportunities in Weather and Climate* published in 1986 went far beyond what could be regarded as the pioneer 1970 volume *The Value of the Weather*. While the latter provided for the first time an overview of the 'economic dimensions' of climate and human activities, it stopped short of addressing the manner in which decisions involving the atmosphere are actually made. *The Uncertainty Business* provided fresh insights into this key issue. Specifically it considered how the variable nature of a particular and elite resource so often taken for granted - the atmosphere - must be accepted as an integral part of the management package.

The current book *The Human Impact of Climate Uncertainty: Weather Infformation, Economic Planning, and Business Management* highlights the main features of *The Uncertainty Business* and is written for the non-specialist. It should be of particular interest to decision-makers in agriculture, economics, and business, students in geography, economics, planning, engineering, and agriculture, as well as the informed and interested layperson and politician. It will also be a useful guidebook for the meteorological professional who needs to become more informed on the economic, social, political, legal, and strategic implications of the past, present, and future weather and climate.*The Human Impact of Climate Uncertainty* consists of eleven chapters.

Chapter I, THE SETTING, first considers the political realities of climate sensitivity. It then looks at the changing ideas associated with the terms 'weather' and 'climate' especially in the operational decision-making area, and considers the various aspects of the weather/climate 'game' as played by its many participants.

The development of the concept that the atmosphere is a variable elite resource is then discussed. The nature of this elite resource is examined from the viewpoint of understanding its utility in human terms, and specific attention is given to the serious nature of the weather and climate game. Climate solutions to problems are then examined as well as a discussion of 'what should be free' and the analysis of the value and costs of weather and climate information. This leads to a review of how weather events and weather information influence the agricultural and energy sectors of the world's commodity markets.

Following this, an examination is made of the political and strategic aspects of econoclimatic analyses, including their impact on national intelligence agencies. This leads to a discussion of the sensitivity of nations to a changing climate - both now and what is likely in the year 2000. The underlying factor of this chapter and indeed the whole book is the conflicts of goals associated with those who must respond to weather and climate events. As an illustration, an operator of a ski resort may welcome additional snow (and a favourable forecast thereof), whereas the operator of a fishing lodge - downstream from the ski resort - would be anything but pleased with such a prospect. In a like manner the announcement of the likelihood of thick fog at an airport may well be greeted with enthusiasm by hotels, restaurants, and rental car companies in the vicinity, but with dismay by airline operators and those needing to travel. That is, gains to some may well be losses to others.

THE CHANGING ECONOMIC CLIMATE is discussed in Chapter II. An examination of the atmospheric part of what is commonly called the 'economic' climate is first examined, with particular reference to food supplies, and the important 'near average' conditions in which most people and nations live. An in-depth examination of the changing definitions and concepts relating to climate, climatology, and meteorology during the past 100 years is then presented. This review includes the reasons why the economic, social, and political aspects of the subject are now receiving so much attention.

The management of national resources takes place within an institutional framework of laws, policies, and administrative arrangements. These are all conditioned by traditions and other cultural attributes which vary considerably from country to country. The chapter addresses the matter of human responses and public policies to the weather and climate, and the management of the atmospheric resource. This is followed by a discussion of specific public (state) educational, and private (commercial) sector responses of several nations to these questions. Finally, the specific national sensitivity of the New Zealand economy to weather and climate events is examined.

Chapter III considers THE GLOBAL SCENE. The international institutional framework is considered with an emphasis on the changing and significant role of the World Meteorological Organization (WMO), and its various Commissions and Expert Working Groups, together with the supporting role of the Food and Agriculture Organization (FAO), the United Nations Environment Programme (UNEP), and the International Council for Scientific Unions (ICSU). Specific attention is focused on the importance of the WMO World Climate Programme (WCP) and the functions of its four parts (data, applications, research, and impacts). The inter-relationship between weather and climate events is then discussed and it is recognized that while not all activities are economic in nature, many decisions have an economic context regardless of the nature of the activity.

The chapter assesses the specific manner in which climate and various activities interact. This is done at various levels ranging from local to international. The important political realities of variations in the atmospheric resource are then examined, including an in-depth analysis of the carbon dioxide/greenhouse gas climate change challenge. The important - and as yet unresolved question - of the ownership of the atmosphere leads into a discussion of the issues relating to the deliberate and inadvertent modification of the atmospheric resource. It includes consideration of the disposal of effluents into the atmosphere including the very important environmental questions concerned with ozone, the acid rain problem, and a 'nuclear winter'. The question of the political realities of any climate change is also reviewed.

THE INFORMATION, COMMODITIES, AND COMMUNICATIONS mix is discussed in Chapter IV. The chapter opens with a discussion of weather and climate information in the context of the wider issue of forecasts, data, and information. The impact of weather and climate may be assessed in *economic* terms. To do this the various data banks linking the parts of the weather- economic mix are examined.

The wider issues of weather and climate *information* provide the theme for the rest of the chapter. The value and costs of both weather and climate information - particularly as they relate to the observation and transmission of basic data - are discussed. In addition, the dissemination of processed meteorological and climatological products (such as maps, forecasts, data sets, and application summaries etc.) to users, by both public and private media, is examined. An important consideration in this regard is the extent to which a public meteorological agency should charge for services which it provides to the public at large and to specific users. The very important subject of marketing and communication is then considered, with particular reference to the opportunities and challenges afforded by the various sectors of the media, including the exciting developments of user based videotex systems.

Chapter V looks at the important questions of WEATHER MONITORING AND FORECASTING. The strategic aspects of monitoring the various spatial and secular dimensions of the global weather and climate are discussed - emphasizing the now current realization within most of the international community of the necessity to understand *both* the weather and the climate. This, it should be noted, is in contrast to the World Weather Watch approach of the 1960's and 1970's when only 'synoptic' weather monitoring was considered to be necessary, such measurements being restricted in the main to supporting purely 'meteorological' activities such as marine and aviation forecasts.

Weather forecasting is next examined, and notes that the public characteristically regard accurate, reliable, and timely forecasts as being the prime product - and in some cases the only product - of a national meteorological service. Further, they believe that forecasts are all that is required to make sound judgements. This, however, is far from the truth. What is needed, in fact, is information about the *past* and the *present*, as well as the *future*. All are equally important, and in some circumstances a knowledge of past and current conditions may be of considerably more value than the forecast conditions, irrespective of the accuracy of such predictions. As an illustration, a forecast that there is a 50% probability of precipitation of a certain magnitude in the next 24 hours is by itself insufficient information for a manager of an electric power utility to decide whether or not to release water from a reservoir. Similarly, such a forecast would be inadequate for a farmer to decide to move stock to a market, or to plant corn instead of wheat. In both of these cases the decision-maker would need to know, among other things, how much water or soil moisture was already available, whether such conditions were normal or otherwise, how such conditions compared with those in 'competitor areas', as well as what the longer term prospects were.

IMPACTS AND SENSITIVITIES are discussed in Chapter VI, which first examines the concept of 'climate sensitivity' as applied to both specific activities and national economies. The challenges of operational decision-making are highlighted. The idea of climatic impacts is then introduced, particularly from the viewpoint - as expressed in the World Climatic Impact Programme - that nations not only can maximize the benefits of correctly utilizing the atmospheric resource, but also can reduce the unavoidable losses associated with its changeable nature.

Chapter VII examines the important concept of WEIGHTING THE WEATHER. An in-depth analysis and explanation of the need to assess weather and climate information in terms of its significance to various areas and populations is given. This type of assessment may be described as the 'weighting' of weather and climate information, in the same manner that a consumer price index 'weights' the various components of the 'consumer basket' according to their importance. Decision-makers such as national producer boards, or energy authorities, now believe that the weighting of weather and climate information is essential to their deliberations. This is especially true in Canada, the United States, and New Zealand. This new approach elevates applied climatology from the classical text book age (still practised in many national meteorological services and educational institutions) to the operational and planning age.

THE AGRICULTURAL SCENE is discussed in Chapter VIII and considers various aspects of the drought problem, pastoral production and weather relationships, and consideration of how weather *and* weather information affects the commodities market, as well as the use of commodity-weighted weather and climate indices at the international level. Specific examples are given of the development and application of 'climatic confidence indices' on both a national and international scale. These latter indices may be considered to be the climatological equivalent of 'leading economic indicators' but *because of their availability in real-time*, they may in certain circumstances also be considered to be predictors of 'leading economic indicators'.

WEATHER RELATIONSHIPS are assessed in Chapter IX, which specifically relates the weather and climate resource to the electric power, manufacturing, retail trade, road construction, transport, and construction sectors. These assessments incorporate the use of suitably weighted weather and climate information, and show clearly that for any time period, what is 'favourable' for one sector is not necessarily 'favourable' for other sectors. The chapter concludes with an examination of New Zealand income and product accounts from a weather and climate viewpoint.

In Chapter X, FORECASTING PRODUCTION is discussed which takes the assessment process one stage further, namely through the use of commodity-weighted weather and climate information in formulating weather/climate based forecasting models of economic activities and production. The various types of prediction models are discussed, followed by specific examples relating to the development and use of weather/climate based forecast models for the dairy and wool industries. In addition, the development and use of weather/climate based economic forecasting models is considered, including their real-time application to national business activity indicators.

THE FUTURE is considered in Chapter XI. First a futuristic agenda for a 15 March 1994 conference is discussed in the context of the 'Weather Administrator's Day'. It suggests that within a decade the meteorologist will be making significant economic decisions with considerable political overtones. The challenging 'information opportunity' provided by the real-time availability of a vast amount of weather and climate data is then discussed. Finally, the potential global and national issues which need to be considered before we reach the 21st century are assessed.

ACKNOWLEDGEMENTS

It is a pleasure to acknowledge the help I have had from many people in the preparation of this book. In particular, a special word of appreciation is extended to the late Dr Derrick Sewell (formerly of the Department of Geography, University of Victoria, Canada) for his suggestion and initial encouragement to me to write a 'lay-person's' version of *The Uncertainty Business:Risks and Opportunities in Weather and Climate* (published by Methuen and Co., London in 1986 and Methuen Inc., New York in 1987), and to Mr Oliver Ashford of Didcot, Oxford, England (formerly of the World Meteorological Organization in Geneva, Switzerland) who critically read all of an earlier draft text. I would also like to thank Dr Martin Parry of the Department of Geography, University of Birmingham, England for his very useful comments on the draft text. I hope that the final text reflects their wise counsel.

Several other people also read parts of the text and I would particularly like to thank my colleagues in the New Zealand Meteorological Service fot their valuable criticism and assistance. Special thanks in this regard are due to Mr J.S.Hickman, Director of the New Zealand Meteorological Service from February 1978 to September 1988, for providing facilities for the preparation of several sections of the book, some parts of which have been previously published in reports issued by the New Zealand Meteorological Service. I also wish to express special thanks to the New Zealand Meteorological Service for the excellent secretarial and technicical support I have had over many years, and in particular to Marco Overdale for introducing me to the 'mysteries' of the *Pagemaker* word processing system, on which the camera-ready pages of this book were prepared.

The very helpful co-operation and guidance of Flo Campbell, Emma Waghorn, and in particular Mary Ann Kernan of Routledge's London office is also very much appreciated.

Acknowledgement is also made to several granting agencies and organizations who directly or indirectly have supported my endeavours in the economic and social aspects of meteorology and climatology over the last 30 years. Special thanks in this regard are due to the New Zealand Meteorological Service, the University of Victoria (Canada), the University of Missouri (United States), the University of Delaware (United States), the World Meteorological Organization (WMO), and the United Nations Environment Programme (UNEP).

I would also like to thank various societies, editors, publishers, organizations, and individuals for the use of, in part or in whole, tables and/or figures, and specific textual quotations. Specific acknowledgement is given at the appropriate place in the text, either by direct reference to the source, or by a footnote keyed to the bibliography.

Finally the inspiration offered to me over many years by the late Jim McQuigg of Columbia, Missouri, the late Stuart Hurnard of the New Zealand Meteorological Service, and the late Derrick Sewell of the University of Victoria (Canada) is gratefully acknowledged. I am sure that without their guiding hands this book would not have been published.

*This book is based in part on **The Uncertainty Business: Risks and Opportunities in Weather and Climate** (Methuen and Co., London, 1986; Methuen Inc., New York, 1987). For more complete acknowledgements, the reader is referred to the Acknowledgements in the earlier book.*

New Zealand Meteorological Service
Wellington
New Zealand

30 September 1988

THE SETTING

A. CLIMATE SENSITIVITY - POLITICAL REALITY

In his review of *Climatic Constraints and Human Activities*,[1] Sir John Mason (formerly Director-General of the British Meteorological Office) stated the following:[2]

> Dear reader, I fear that a new multidimensional, multidisciplinary, cross-cutting subject of the utmost importance and concern, namely socioeconomic climatology, is about to be launched on an undeserving world and, if this book is a foretaste of things to come, let it be a warning.

These comments highlight the difficulties facing economic climatology, and the differing viewpoints of individuals. Indeed in reviewing the same book I took quite a different viewpoint.[3]

> This is an important book because it is concerned with what many would regard as the non-scientific aspects of weather and climate. ... But let us not deceive ourselves in thinking that all is easy when the non-scientific aspects of the subject are considered. It is not, and terms like education, marketing, awareness, relevance, and value are often much more difficult for both meteorologists and users of meteorological information to come to terms with, than the more familiar terms - at least to most meteorologists - of vorticity, divergence, convection, radiative processes etc.

The real truth probably lies somewhere between these views and it is appropriate to quote Sir Brian Pippard, Professor of Physics at Cambridge University, who in giving an inaugural lecture[4] at the 150th anniversary meeting of the British Association in 1981 questioned whether methods applied in the laboratory by the physicist, biologist, chemist, and mathematician have any potential value for the politician and sociologist in resolving problems. He illustrated the connection between the problems of weather forecasting and the difficulties in making political predictions with a series of models showing the instabilities that cause earthquakes, electronic interference, and the disorderly flow of liquids and gases. These he said were examples of events leading to chaotic behaviour, and he wondered if lessons could be drawn.

Attention is focused in this book on the difficulties and challenges associated with the mix of the economic and meteorological systems. In defining and measuring weather and climate sensitivity it is important that the applied nature of the problem be emphasized. In particular, in considering weather and climate sensitivity and the linkages with political reality, it is essential that the politician and the planner become more weather and climate orientated. Central to this planning is the need for a much more comprehensive monitoring and analysis of the world's climate, both to detect and to predict changes. The sensitivity - in economic, social, and political terms - of nations to weather and climate variations must also be better understood.

B. WEATHER AND CLIMATE: IS THERE A DIFFERENCE?

Television, radio, newspaper, and business journal reports during the 1970's made many people aware of a succession of climatic episodes that had worldwide economic, political, and social repercussions. For example, the Sahel drought of 1970-75 brought widespread famine, and the freezing weather and drought in the Soviet Union in 1972 led to large grain purchases abroad.

These developments point to the changing viewpoint of what, in the English language, is meant by 'weather' and 'climate'. Traditionally, in terms of time, 'weather' has been regarded as the events that happen over a short period, usually a few hours to a few days, whereas 'climate' has been regarded as the events that happen (or happened) over long periods, usually years to decades to centuries. Unfortunately, the English language does not have a word which describes the important recent past (in terms of weather and climate) and perhaps English should adopt the German word 'Witterung', or the Japanese word 'tenkoo'.

Politically it is the recent past and the near future that are most important. Hence not surprisingly the Assessment and Information Services Center of the U.S. National Oceanic and Atmospheric Administration (NOAA) during the early to mid-1980's issued 'climate' impact assessments on a *weekly* basis. Although some would argue that they were really 'weather' impact assessments, their important feature is that they discussed an important period of time (i.e. a week) as far as the political, economic, and social life of the United States was concerned. The word 'weather' should still be used to describe what happened over short periods of time - but the time periods generally accepted are now much more of the order of days, rather than weeks or months.

But, the *real* economic and political value of weather and/or climate information is realized only when it affects the way decisions are made. To assist decision-makers, national meteorological services and several private meteorological companies prepare and disseminate a wide variety of weather and climate forecasts - ranging in time from a few hours to a few months.[5] In addition, a few national meteorological services and meteorological consultants prepare and disseminate commodity-weighted weather and climate information (that is information related to people, production, consumption rather than areas) in real-time.

Irrespective of how the terms 'weather' and 'climate' are defined, the overall impact of *short-term* variations in these parameters will continue to be important. Paramount to any discussion of weather and climate variations is therefore a better understanding of their impacts on communities and nations. Despite this, many weather and climate related activities have until recently been planned as if the climate was stable. During the 1972-75 period, however, the world was made very much aware of the catastrophic drought in the Sahel area of Africa, with all its political and social implications. A decade later, similar weather and climate impacts are still occurring, as is evident by the disastrous drought in Ethiopia in 1983-85, and the impact of the 'El Nino' in many parts of the world in 1982-83 and the summer drought and 'heat wave' in the United States in 1988.

C. UNDERSTANDING THE ATMOSPHERIC RESOURCE

1. The Setting

The atmosphere is an 'elite' and special type of resource which may be tapped, modified, despoiled, or ignored. Most societies also forecast the availability of the various components of this resource, primarily based on knowledge of the nature of the atmosphere through observations made and collected globally in real-time. To use these forecasts correctly requires information not only on the past weather and climate, but also on knowledge about the inter-relationships between the social, political, and economic actions of societies and individuals, and atmospheric events. Together, this package of information assists in developing the ability of people to manage this elite resource, thereby improving the economic, political, and social outcome of the many weather-sensitive activities.

The international community, through United Nations Conferences on Food, Water, Population, Energy, and Climate, and many nations have in recent years accepted the view that there are physical limits to the availability of natural resources, including the atmosphere. One of the major problems, however, is the recognition by people that the atmosphere is an elite and variable resource. This variability can be evaluated in ways useful to both political and economic planning and management. Of prime importance is the *impact* of short-term variations, and these impacts are expected to continue to increase in importance as the demand for food and water increases, and the relative cost of energy increases.

It is often argued that information about the atmospheric resource cannot be used in economic, agricultural, and political planning until accurate climate forecasts are available. It must be appreciated, however, that most economic planning, and most decisions involving climate sensitive activities are made on the basis of the information (including forecasts) actually available, and *not* on what 'could be' available. Clearly, planning decisions have to be made, and qualified meteorologists and climatologists need to provide the best possible guidance to decision-makers. One of the functions of many national meteorological services is to apply meteorological science for the 'benefit of the community'. Included in this function are forecasts but an equally important function is the provision of *information* about the weather and the climate.

2. Information and Decision-Making

People, firms, and governments make many weather and climate-related decisions. Such decisions are based on the information available. Accordingly, whenever such information is used in decision-making such as in connection with forecasting Australia's wool production, the world's coffee price, or Canada's heating energy requirements, it is essential that the weather and climate information be in a form appropriate to the problem and the area concerned.

The cost of providing such information is of course important, it is really quite low especially when there is the capability of using this information in a deliberate manner both to monitor and control an economy. For instance, the analysis of the weather of one country by another, so as to forecast agricultural production or energy consumption in the competitor country, is a reality. Close monitoring of the atmospheric resource is therefore fundamental, and will continue to have important economic, social, and political consequences in many countries.

An increasing world population and a desire to raise living standards have increased the pressure on the natural resources of food, water, and energy. It is important therefore that planning authorities recognize the variable nature of weather and climate, and that any change may have an even greater economic and social impact in the future, than they have had in the past.

National meteorological services endeavour to provide accurate and timely weather and climate information to decision- makers, but this information is much more than tomorrow's forecast or the average January temperature. Indeed, many people are convinced that the real-time availability of weather and climate information is of considerable benefit in providing more relevant answers to the vagaries of the 'economic climate'. For example, the atmospheric resource is fundamental to the agricultural competitiveness of many countries, and to gain maximum benefit agricultural decision-makers need to be provided with *real-time* weather and climate information. Such information should be 'weighted' by economic activities and areas, which is available on a timescale which provides sufficient lead-time for decision-making.

D. IMPORTANCE OF WEATHER AND CLIMATE VARIATIONS

Economically, socially, and politically the world is becoming more sensitive to the resources of the atmosphere, and any variation in these resources have now become a significant factor in both short and long-term planning. For example,

Newsweek (11 April 1983) in noting the importance of weather on produce prices in the United States stated:

> Produce markets in the East and Midwest are already suffering, too. 'I can't remember a year with weather like this,' says A.H. Nagelberg, president of the New York Produce Trade Association. 'Wholesale prices are ridiculous.'

In a like manner *Business Week* (31 January 1977) in an article 'An Old-Style Winter Disrupts the Economy' said that '... the most critial impact of the frigid weather is sharp curtailments of natural gas, which have resulted in widespread shutdowns of industrial plants'.

The potential economic, social, and political value of past, present, and future weather is also made very clear in an article on 'Drought in Africa' in *The Economist* (11 June 1977). Under the title 'Acts of Man', the report stated:

> The Sahelian disaster prompted a guilty realisation that a new assessment of the underlying causes of drought was necessary to prevent it ever having such consequences again. ... A succession of good years disguised the incipient calamity. Only a temporary deterioration in the climate was required to bring the whole system to its knees.

More than two decades ago, when food surpluses were probably at their highest, Watson[6] stated:

> Climate determines what the farmer can grow; weather influences the annual yield, and hence the farmer's profit, and more important especially in under-developed and over-populated countries, how much food there is to eat.

In the light of recent climatic and economic events (such as the Sahelian droughts, and the El Nino/fisheries/tropical cyclones/harvest failures inter-relationships), the value of having and using information about the past, present, and future weather and climate should be accepted without question.

E. WEATHER AND CLIMATE: A SERIOUS GAME

Over the past few decades many aspects of the atmospheric environment have been examined, but only a fraction are concerned with what the *public* considers to be weather. Within each facet there are a variety of specialists and generalists and although there is some cross-fertilization, most meteorologists and climatologists tend to become specialists. The intricacies of what the author called 'The Weather Game' are described in a Presidential Address to the Meteorological Society of New Zealand[7] which looked at many different aspects of meteorology and climatology. Each aspect was considered in terms of its overall importance, application, impact, political significance, and possible role in the international scene. The topics discussed can be looked at in the form of a game - with its specific rules, protocol, history, adherents, and sceptics, but like all real games, one must play the game to achieve satisfaction, and also be part of its 'administration' (or otherwise be very influential) if one hopes to change any of the rules by which the game is governed.

The value of meteorological and climatological information is now recognized as an essential ingredient to the weather/climate game, as well as the need to consider the urgent problems concerned with the limited atmospheric resources. But much more could and should be done in this regard, and this book endeavours to provide the vision and lateral thinking that is (and will be) necessary if meteorology and climatology are to become really involved in the ultimate 'uncertainty business'.

F. CLIMATIC SOLUTIONS TO PROBLEMS

A key aspect of climate and socio-economics is asking the questions 'what is your problem' and 'is there a weather or climate solution' to your problem. In the case of agriculture one can identify the biological response to a variation in the environment; for example, very dry conditions may decrease wool production but increase wool quality, while cooler conditions may decrease corn production but increase potato production. The significant question then - given this information - is: 'what does the farmer, farm advisor, agricultural producer board official, or government Secretary of Agriculture do?' As shown in Fig. I.1 one can either do nothing and accept the economic, social, and political consequences, or one can act.

Many options of 'doing something' are available, including relocation to a more suitable area (which may be the only choice if a longer term climate change such as a greenhouse gas warming is envisaged), or changing the method of operation by using for example a different species of plant. One can also stop production (such as deciding not to plant wheat or rice), modify the environment by growing a shelter belt of trees, provide irrigation water, or deliberately modify the weather. In all cases, the choice of doing something should improve production and/or quality, and allow more useful management decisions to be made. The role of managers, with their differences in willingness and ability to assess risk is crucial in many of these options, but as vividly described in *Time* (28 March 1983) the situation in many cases is complex and extremely difficult.

Even the region's hardiest inhabitants find the eerie silence unnerving. Under the blazing sun, the quiet that grips vast stretches of eastern Australia is accentuated by the whine of the hot, blast-furnace-like winds that bear down on farmlands long since stripped of vegetation, whipping up the bone-dry topsoil into whirling clouds of reddish-brown dust. Here and there, the superheated

Figure I.1 Patterns of responses leading from environmentally induced production events to a management decision

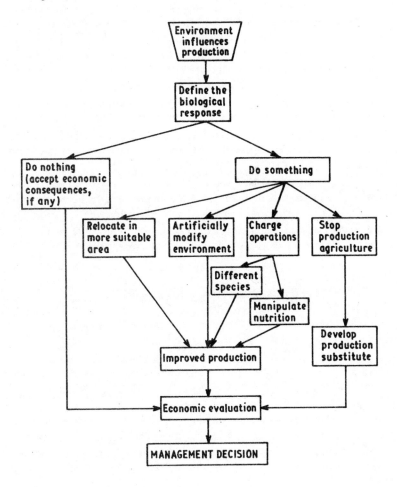

Source: After Maunder (1985), from National Academy of Sciences (1976)

stillness is broken by the squawking of white cockatoos and black crows, wheeling and circling overhead, searching the parched land for dead and dying animals below.

Droughts such as those described above of course occur in many other areas, and in a study of a much less severe drought situation in New Zealand - written as it was happening[8]-the kinds of 'normal' decision-making that faced one farmer at that time are pertinent. The particular farm of 210 hectares had increased its sheep population to 2,500 by January 1982, but because of the drought $NZ30,000 had to be spent in providing supplementary feed. This included $NZ17,000 on barley which was brought in as a grain feed to carry the ewes over the winter (June - August). The farmer paid another $NZ6,000 to graze sheep on a nearby farm, and the balance of $NZ7,000 went in 'invisible' costs like selling lambs at low prices. The farmer in this case was an established farmer and he was able to carry these costs. As a result of his decision to purchase feed, he was able to close off most of the farm over the winter and as of September 1982, with the drought reaching a peak, the pastures were in reasonably good condition and the ewes (which were fed on barley) had gone into lambing in good condition. A key question is 'what would have happened if $NZ23,000 had not been spent on supplementary feed?' and further 'now that $NZ23,000 has been spent, what is the correct strategy if the drought continues?'

The comments relating to this farm show that decisions have to be made - even the decision to do nothing. Moreover, until a reliable long-range weather or climate forecast is available, a careful analysis of the climatological record and its probabilities - together with appropriate advice from agricultural and financial advisors - may well provide the only real guide to the correct strategy.

G. FOOD, WEATHER INFORMATION, AND POLITICS

The growth in the world population requires production increases of almost 30 million tonnes in food and feed grains per year.[9] To achieve this goal a concerted global approach is needed to monitor the tightening food supply/demand situation, and information provided by agricultural meteorologists is clearly very useful in providing more lead time for planning and decision-making at the national and international level.

The use of both the vast amount of climate data in the archives of national weather services, and the real-time weather information available over the WMO Global Telecommunication System[10] presents opportunities to agrometeorologists at the international, regional, and local levels. These include the provision of summarized past and present weather and climate data to the FAO Global Information and Early Warning System on Food and Agriculture, in association with the monitoring activities of the WMO World Climate Programme; the weather-based prediction of agricultural production prospects as a basis for planning agri-business operations, export/import policies, and price support controls; as well as the additional interpretation of weather and climate information in terms useful to economists, scientists, extension officers, and farmers.

There will always be competitive advantages to those who gain first access to and know how to use such information, and in certain circumstances weather information can become a very powerful tool. Indeed, how long a more publicly accessible system for predicting agricultural production from the already available weather and climate data of the World Weather Watch system can remain 'suppressed' is debatable. The continued present 'free' exchange of the world's weather and climate information may also be in doubt if the use of weather and climate information, particularly in relation to its use as an early predictor of agricultural production, becomes too political, or governments become too 'user-pay' orientated at the national and local levels.

H. POLITICAL AND STRATEGIC ECONOCLIMATOLOGY

Politicians have to look at an extremely wide range of possible responses to their policies and their decisions, which are primarily based on political goals, view such things as climate in combination with other issues. In discussing this situation in Canada, Sewell and MacDonald-McGee[11] suggested that inevitably governments will become increasingly involved in the management of climate resources, but that it is unclear whether present legislation, policies, and administrative structures and procedures are adequate to meet this challenge.

In an associated manner, Meyer-Abich[12] in a paper 'chalk on the white line' discussed the transformation of climatological facts into political facts and commented:

Approached by a climatologist who is reporting his latest results in terms of temperature, pressure, and humidity, the politician will ask: 'All this may be so, but what difference does it make to me? I am concerned with achieving goals in a given situation, and how does the climate come in here?' Obviously the climatologist cannot answer this question within the framework of his scientific discipline. He may, however, hand over the question to the geographer and to the ecologist, since these researchers are concerned with the habitat of mankind and other species, and can interpret climatological parameters through their disciplines as impacts on the conditions of life.

The writer's comment that the 'climatologist cannot answer the question' is perhaps a little unfair for there are climatologists who can answer these questions. Nevertheless, the point is generally valid, and should give food for thought to many conservative-thinking climatologists.

Chapter II

THE CHANGING ECONOMIC CLIMATE

A. THE REAL ECONOMIC CLIMATE

1. Area, Time, and Business Aspects

The phrase 'economic climate' is a favourite expression amongst politicians, economists, company directors, and the news media. But what is the true 'economic climate' of a nation? Websters's dictionary gives one definition of climate as: 'the prevailing temper or environmental conditions characterizing a group or period', which suggests that the *'true'* economic climate involves consideration of not only the political, social, and economic life of a nation, but also environmental factors which should include the weather and the climate.

For example, the *Financial Times* reported on 14 December 1985 that coffee prices had been rising virtually without interruption all that week on the London and New York futures markets in anticipation of severe drought damage next year to the Brazilian crop, with the London futures market closing at £2,148 pounds per tonne, upon the week by £99 per tonne, its highest level for six months. But few substantial studies have been completed on the specific effects of weather on business. Indeed it is paradoxical that the value of information about the weather and climate to business and commercial activities appears to be ignored, not only by many meteorologists and climatologists, but also by the business community. The relatively flourishing consulting activities performed by meteorologists in the United States and a few other countries are exceptions; indeed, they provide clear evidence that *some* astute business enterprises are willing to pay good money to know more about how weather and climate influences their business.

The value of weather and climate to an area requires the identification of activities that are affected directly or indirectly by the weather and climate, and an analysis of the manner in which a specific change results in gains or losses to such activities. The problems of evaluating weather in an area have been considered by a number of investigators but in most cases these (and most latter) investigations[1] either have been spatially restricted, or they have been related to 'long periods' of at least a month. Such studies cannot therefore contribute to answering the question: 'What is the value of the weather to a large area (such as the United States, France, or Argentina) over a short period of time (such as a week)?" However, in an analysis by the author on retail trade,[2] a 'large area' and a 'short time period' *was* deliberately used in a study of the United States weekly retail trade. These factors were chosen for two reasons: first, that variations in nationwide economic activities are important to various high-level decision-makers, and, second, only weather over a short period has any real practical meaning to the millions of low-level decision-makers who go about their daily and weekly activities.

It may be argued that there is little point in such analyses, since any relationships that might be derived cannot be used because *accurate* weekly weather forecasts are not yet available. However, such forecasts *will* become available, and decisions based on such forecasts will be made. For example, five-day forecasts are issued on a regular basis in many countries and are widely distributed by the news media. Of course, it does not necessarily follow that business decisions are actually based on the effect that the weather has or will have on a particular business, but what if business (including business on a national scale) chooses to ignore weather information (past, present, and future). Should it be part of the meteorologist's responsibility to educate the business community to utilize better such information? Further, if meteorologists are *really* serious about forecasting the weather, then one justification for even more effort in this direction is the potential use of this information by a whole host of producers, and consumers, from all areas of the community.

Traditionally, national meteorological services have observed, collected, and processed synoptic weather information in real-time. Recently, in a few countries such information has also been analysed in an *applied* sense in real-time, so that key econoclimatic indices are now available regionally and nationally, as well as on an international basis. 'Climate productivity indices' and 'international weather indices' for key commodities are therefore a reality. Furthermore, the availability of such information through videotex type systems requires an awareness by both meteorologists and climatologists of the marketing potential of their services, and also by the decision-makers who can use such information.

2. Information, Economics, and Politics

Meteorologists and climatologists need, and should welcome, the help of economists and others in assessing the likely economic and social impact of climatic changes but this report of a task force convened by the International Institute of Applied Systems Analysis (IIASA) in Vienna does not inspire confidence.

These comments by Mason[3] relate to a review of the book *Climatic Constraints and Human Activities*[4] and both the book and Mason's critical review provide much food for thought. As noted in Chapter I, the real economic climate *is* important and one may ask why the comments on economic meteorology by an eminent meteorologist like Sir John Mason[5] are often met with scepticism by some sections of the meteorological community, yet his writings on cloud physics are highly respected. Perhaps the answer lies in the fact that we all think it is easy to do a better job than a social scientist, economist, or politician but much more difficult to do a better job than a physical scientist.

The 'other side of the meteorological fence' can in fact be quite different but the influence of the relatively few exponents of socio-economic climatology is far greater than their numbers would indicate. For example, it is very evident from the reading of any submission for funds (to government, granting agencies, etc.), or any annual report of activities, that the economic, social, strategic, and political justification of the subject is now always given prominence.

It is important that applied climatologists provide appropriate guidance to economic forecasting groups. Much of the necessary thinking for this guidance involves areas peripheral to the normal activities of climatologists and economists, but the problems are real, and regrettably it is the 'atmospheric' component of an economy that is most often ignored. A key factor in meteorologists and climatologists providing this guidance is the availability of weather and climate data in real or near real-time, and the international meteorological system allows such data to be both collected and processed in real-time, as well as the potential to produce - also in real-time - at least provisional estimates of economic and social impacts.

One specific aspect of these new products is their use in weather-adjusting national economic indicators, so that the true economic climate of a nation may be ascertained. For example,[6] it is possible to 'adjust' the *Business Week* 'Index of United States weekly economic activity' (which uses weather-sensitive electricity consumption as one of several key components) so that it more truly reflects 'economic strength' rather than 'environmental strength'. Similarly, it is possible to 'adjust' consumer price indexes so that they are more clearly adjusted for the influence of abnormal weather on prices.

The challenge facing the weather forecaster, the weather services expert, and the climatologist, is to produce practical information that can be readily understood and integrated into an operational and planning process. The chances of success are improved when the operation and planning process is well understood - they are much improved when the user is convinced, involved, and pays for the information.

3. Weather and Climate and Human Well-Being

The journal *Newsweek* in a major story in its 9 May 1983 issue on 'Africa: The Drought this Time' quotes a district administrator in Ethiopia: 'We have no seeds because we have already eaten them. ... If the rain falls, people will need seed and oxen, but they don't have either. ... Worst of all, of course, they don't have rain'. Then, in an *FAO Special Report on Foodcrops and Shortages* issued on 15 April 1983 it is stated:

... In six West African countries crop failures towards the end of 1982 due to drought have resulted in a tight domestic food supply situation. ... Recent rains in Australia may have broken the drought, this improving prospects for planting the 1983 wheat and coarse grain crops'.

Such quotations clearly point to *one* significant aspect of climate and socio-economics, namely the dependence of *people* on favourable weather and climate conditions. But there is much more to the subject than these events, however distressful they are to the people and governments concerned. For example, although last week's weather is history, the measurement of its economic and social impacts - using conventional economic techniques - is not known until several weeks or months after the event. However, it is quite feasible using appropriate real-time weather and climate data to make a forecast of economic and social *data*, which at the time of making the forecast either *do not exist,* or have yet to be compiled. This is one reason why the real-time climate and socio-economic mix is so important, a reason which has significant political overtones.

B. CHANGING DIRECTIONS IN CLIMATOLOGY

1. An Overview

Climatology has, until recent years, been mainly considered outside the central concern of meteorologists. In addition, within geography, climatology is nearly always associated with *physical* geography, despite its strong linkages to people and economic activities. Moreover, the public has traditionally thought of climate only in terms of averages with the result that political behaviour often ignored long-term changes or the occurrence of significant climate events such as droughts in Africa. Recently, a real appreciation of climate variability has grown in the scientific community - but is only slowly being recognized in the political and economic communities - and a major effort in what might be called 'meteorological climatology' has been initiated at both the national and international level. The following sections review the changing directions in climatology - past, present, and future.

2. Climate Data

Initially, most climate data were used to compile 'statistics' of the climate of places. Later, these 'statistics' became part of the climate paper archive, and more recently have formed the basis of computer data banks. This first use of climate data is fundamental to most studies of the climate, and with the current emphasis on global climate monitoring in near real-time, these data are essential for placing the present climate in its correct 'historical' setting. Education was the second use of climate data, as is evident from looking at any of the 'classic' textbooks on climatology published in the first four decades of this century. The education element initially relied heavily on the availability of the climate archive - which in those days was mainly monthly averages of the various climate elements.

The third use of climate data developed mainly from the 1930's when it became more generally realized that the climate *did* vary, and that the climate archive could be used for planning in the medium and long term; for example, in the design of bridges to withstand extreme wind gusts, and in the optimal economic use of the land/climate resource in designing cropping systems. The fourth use of climate data was for research; indeed until the late 1960's research on climate was with few notable exceptions confined to a traditional and mainly descriptive role. The World Climate Conference in 1979 focused attention on a 'new look' *climate* research programme with the emphasis on climate research and the use of climate data and information in operational decision-making. Many of these applications involve the monitoring of the climate in real-time and they provide information for the up-market decision-maker.

The use of climate information in impact studies is the latest development and was recognized at an international level at the 1979 World Climate Conference. Subsequently, the World Meteorological Organization (WMO) approved the World Climatic Impact Studies Programme in which the United Nations Environment Programme (UNEP) is the lead agency.

3. The Development Phase

Climatology in its earliest meaning was the study of the world's 'normal' weather conditions as varying with latitude and season. However, over the centuries climate and climatology gradually achieved a new meaning and came to mean the *synthesis* of the atmospheric conditions over a region or at a place as influenced by all environmental factors. By the beginning of this century, however, the words 'climate' and 'climatology' at least among British meteorologists came to imply only the study of the average, or rather the most probable weather conditions. The climate of places and areas was simply obtained by statistically treating meteorological data taken over the available network of observing places. But there was an important distinction between the viewpoint of the French and Russian schools, and the English language school over the definition of climate.[7] The French and Russian view was that climate is essentially 'a synthesis of the atmospheric conditions emanating from an interaction between the air and other environmental factors.' More importantly the French and Russian schools considered climatology as the *overall* scientific subject in which meteorology forms only that part which deals with the physical processes in the atmosphere. In contrast, climatology, from the viewpoint of meteorologists of the English language school, became only a *part* (and usually a minor part) of the overall subject of meteorology.

Developments in climatology during the 1900-50 period were highlighted by the work of several well-known climatologists. Among these developments were those pioneered by Thornthwaite[8] who used the term 'topoclimatology' in calling for what he saw as a *new* climatology, separate from the immediate concerns of the meteorologist and rid of the idea of average weather. To Thornthwaite, this was a climatology focused upon the earth's surface and concerned with the exchanges of heat and water at the 'bottom' of the meteorologist's atmosphere. Thus, he saw an active, positive role for geographically orientated climatologists quite different from their more accepted role as 'the keepers of the records'.

Another noted climatologist of this period (and later) was Lamb, who effectively summarized the pre-1950 situation in which the *static* nature of climate was to the fore.[9] He noted that the conclusion that 'climate is essentially constant' was at odds with the acquired wisdom and experience of previous generations but it was a convenient one for those practical operations using climatic statistics for planning. Further it meant that valuable statistical techniques could be developed to derive estimates of the extremes of temperature, etc. and the average recurrence intervals of rare events. Climatology therefore became, according to Lamb, essentially the bookkeeping branch of meteorology - no more and no less.

4. 1960-70 Changes and Socio-Economic Developments

Many changes to the climatological scene took place in the 1960-70 period particularly in regard to the traditional view-point of a constant climate. For example, Smith[10] in 1961 considered the *significance* of climate variations in their association with agricultural production and stated that if recent climatic variations are examined through the eyes of a statistician, we may be forced to conclude that not one of them is 'significant'. He further noted that 'while this may be acceptable to the pure scientist, it could tend to infuriate the man who has experienced what *appear to him* (my italics) to be a radical change in the circumstances in which he has to live and work'.

Another feature of the climatological scene in the 1960's was the beginning of studies linking weather and climate activities to socio-economic activities. But the idea that an economist or a sociologist could be usefully employed in a national meteorological service was still far from reality.

While the 1960's brought significant changes to the climatological scene, it was not until the 1970's that there was general recognition that climate not only is a resource but also can have a profound influence on the economic and political scene. The reasons for this recognition are noteworthy: first, McQuigg[11] wrote a Ph.D. thesis on *The Economic Value of Weather Information*, the United States Weather Bureau prepared a report on *The National Research Effort on Improved Weather Designation and Prediction for Social and Economic Purposes*,[12] and the Australian Bureau of Meteorology called a conference on *What is Weather Worth?*[13] A report by Thompson[14] on *The Potential Economic and Associated Values of the World Weather Watch* was also written at this time, and the author wrote a Ph.D. thesis on *Assessing the Significance of Climatic Variations on New Zealand's Agricultural Incomes*.[15]

A further significant development occurred in 1965 when a symposium con-vened in the United States on *The Economic and Social Aspects of Weather Modification* resulted in the publication *Human Dimensions of Weather Modification*.[16] Participants in this symposium including economists, sociologists, biologists, geographers, political scientists, planners, lawyers, and climatologists. Later, in 1970 the author's first book *The Value of the Weather*[17] was published, which focused specific attention on the 'value' of our atmospheric environment. This new focus of attention was given considerable emphasis during the early 1970's and in 1976 the Executive Committee of the World Meteorological Organization issued a *Statement on Climatic Change* which said:

> It is important to emphasize that information regarding the impact of climate variability on human activities is essential ... in the decision-making process. The methodology to be developed ... should aim at making it possible to present ... the impact of climate variability in terms of production figures, costs, or other similar measures which can be used directly by the economists, planners and politicians.

5. National, Regional, and International Programmes

The mid-1970's continued to focus attention on the socio-economic implications of weather and climate, with the UN specialized agencies being well to the forefront, with World Conferences on Food, Population, and Energy. Later (in 1979) WMO organized the World Climate Conference. There were also important contributions to the 'non-scientific' aspects of meteorology and climatology - notably by Scheider in *The Genesis Strategy*[18] (a critical commentary on climate and global survival) and by Glantz, who is one of the few social scientists seriously involved in the social aspects of weather and climate.[19] The work on climate and socio-economics of the Canadian Climate Centre,[20] the Oklahoma Climatological Survey,[21] the Assessment and Information Services Center of the U.S. National Oceanic and Atmospheric Administration,[22] the Climatic Research Unit of the University of East Anglia,[23] which has promoted an awareness of the effects of climatic variations on people and society, the International Institute for Applied Systems Analysis,[24] and the SCOPE volume on *Climate Impact Assessment: Studies of the Interaction of Climate and Society*[25] should also be noted.

The various parts of the World Climate Programme are closely associated with economic and social matters. However, it is very easy for *any* programme to lose sight of its goals, and White[26] rightly and forcefully commented:

> The World Climate Program should support a component to analyse the economic and social consequences of climatic events. The problem is difficult. Understanding such consequences takes the program into non-scientific realms. But if the World Climate Program is to be of utility to

Governments and others who must make decisions, then it seems to me it must provide information about social and economic consequences. It has to answer the question 'So what?'

6. The 1980's

The increased activities and concerns of the 1970's continued into the 1980's. For example, in Canada the Canadian Climate Centre was designated as the lead agency for the Canadian Climate Programme, one of its prime aims being to promote a much closer association between traditional climatology (e.g. data management, quality control, archiving services), application climatology (e.g. applications, impact studies, hydrometeorology), weather and climate monitoring, and weather and climate prediction.

Considerable concern was also expressed over the future impact of the increasing amount of carbon dioxide and other greenhouse gases in the atmosphere (see also Chapter III), and Hare[27] in discussing the Canadian situation in 1980 noted that while these increases do not imply great disruption for the Canadian economy, they do argue for readiness, and a determination to make good use of them.

Other developments were the much greater appreciation of the importance of weather and climate information in decision-making, the 'operational' activities associated with the World Climate Programme, and the greater awareness of the need to monitor the weather and the climate on both a daily and historical time-frame basis. In addition, there has been continuing activity and concern over the climatic and environmental consequences of acid rain, ozone depletion, and nuclear war.

7. Directions for the Future

One of the goals for climatology is the development of a climate forecasting capability, and although 'useful' forecasts several seasons in advance may not be feasible in the next decade, any work that leads to an improved understanding of the physical processes governing climate, or the role of surface or human influences on climatic variability, will be valuable. To achieve this an improved data collection and analysis system is needed, together with better quantitative models, and a wide range of applied climate studies.

But climate studies must not be confined to the analysis of atmospheric data: rather, such studies must also consider the human problems of food, health, energy, and well-being. The economic, social, human, and political implications of the climatic understanding developed, need to be applied at all levels: locally, nationally, regionally, and internationally. Meteorologists and climatologists will then be able to make a much more positive contribution to decision-making in both developing and developed countries, whatever their political systems.

C. HUMAN RESPONSES TO ATMOSPHERIC RESOURCES

1. The Media Speak

Information about the weather and climate can play an important role in the decision-making processes associated with management of weather-sensitive activities and enterprises. For example, in a cover page article 'The Uncertain Earth' the U.S. Conference Board magazine *Across the Board* (June 1977) stated: 'Official concern is understandable. As many of us had to relearn last winter, our way of life is predicted on the favors of the weather. ... By one estimate, there is about $270 billion's worth of weather-sensitive industry in this country alone.'

Similar media comments from a variety of business journals since the early 1970's provide clear evidence that the world is *not* becoming less dependent on the weather and climate as people and nations become more 'developed'. Indeed the comment in *Time* (31 August 1981) on 'trouble down on the farm' although referring specifically to the Soviet Union could very easily have been applied to other nations and for other periods. *Time* in fact, at that time, neatly summarized the often precarious position that people find themselves in when trying to live within their climatic income when they commented that the '... forecast is bleak this summer in the *kolkhozy* (collective farms) and *sovkhozy* (state farms) of the Soviet grain belt, where capricious weather has caused a third consecutive bad harvest - with an anticipated shortfall of 51 million metric tons in Soviet grain production.'

2. Atmospheric Interactions and Decision-Making

The need for better and more useful information about the atmosphere has come about as a result of two important concepts: (1) that *information* concerning the atmosphere such as the past, present, and future weather and climate is an important resource, and (2) that given an understanding of physical-biological-sociological interactions with the atmosphere, and given sufficient *information* about atmospheric events, people can at times use their management ability to improve the economic and social outcome of many weather- sensitive activities.

An increasing awareness that *information* about the weather and climate can play a very important role in decision-making processes is evident. Included in these processes are aspects of *national* productivity such as retail trade, rice production, and electric power consumption which are often the concern of high-level decision-makers such as ministers (or secretaries) of governments, or directors of national companies. But can a useful relationship be obtained between weather indicators and economic activities over national space scales? The answer is not always evident, but the publication of weekly and monthly indicators of national economic activity in business journals such as *Business Week* and *The Economist* provides some evidence that there is a demand for economic data on this scale; moreover, since some decisions are presumably based on this information, it appears useful also to incorporate into the decision-making process aspects of the weather and the climate which have an influence on the economic indicators. These inter-relationships are further discussed in Chapter X.

D. SPECIFIC NATIONAL REPONSES : THE ROLE OF THE PUBLIC, EDUCATION, AND PRIVATE SECTORS

1. National Business and Government

Many decision-makers (in both the government and private sectors) appear to be unaware of the potential value of weather information (past, present, and future). Important questions arise from these matters, notably: (1) what is the role of the meteorologist in national economic planning and in advising top-level decision-makers, (2) what is the function of a government meteorologist or national meteorological service, and (3) what is the function of a consulting (or private) meteorologist or company in these processes?

The answers to these questions will vary from country to country, particularly where both government and private meteorologists are active. In the United States, for example, more than 200 professional meteorologists and climatologists are employed in over 30 consulting meteorological companies. Hence the role of the United States 'public service' meteorologist/ climatologist may be quite different from that in countries where 'private' meteorologists/climatologists are few or non-existent. But in many countries (including Sweden, Canada, the United Kingdom, and New Zealand) the traditional 'public service' role of the government meteorologist/climatologist is undergoing a considerable change (especially when 'user-pay' concepts come to the fore). In this regard, the comments by Epstein on the United States situation in the 1970's are pertinent.[28] He said '...the (private) meteorologist represents an indispensable link in our ability to bring the best meteorological services to the nation. We have important jobs to do, and we do them best when we work together.'

2. The New Zealand Approach

The current (as at September 1988) functions of the New Zealand Meteorological Service are set out in the *Civil Aviation Act, 1964*, which states that the Service is to provide a meteorological service for the benefit of all sections of the community, to promote the advancement of the science of meteorology, and to advise the minister and government departments on all matters relating to meteorology. But what are the specific obligations of and opportunities for meteorologists and climatologists in these matters? This is only partly evident from the Act, but New Zealand's Minister of Science in 1976 (the Hon. L.W. Gandar) in speaking about scientists in general said that scientists have an obligation to keep the public informed of new developments in research and their possible impact on daily lives. He further commented[29] that '... the public, scientists, and politicians can choose well only if we are well informed about scientific research and we are made aware of the implications in its applications.'

It may be inferred from these comments and the implications of the New Zealand *Civil Aviation Act, 1964*, (and similar 'Acts of Parliament' in other countries) that meteorologists and climatologists should become more involved in advising top-level decision-makers and assisting in national economic planning. Indeed the alternative is that weather and climate information will be considered only in an ivory tower atmosphere, or left to be dispersed by 'TV weather girls', 'disc jockeys' or journalists. Clearly, weather and climate information is far too valuable to be treated in *only* those ways.

At the party political level in New Zealand, there has - at least in recent times - never been any official policy on meteorology and/or climatology, and certainly little if any reference was made to the *atmospheric* economic climate in the election manifestos of any of the major parties in recent elections. However, the remarks by the (then) Director of the New Zealand Meteorological Service (J.S Hickman) in an editorial in the August 1982 issue of *Weather and Climate* are, in the absence of any official government policy or party political policy, worth noting. He stated:

> ... We have a previously undreamt of opportunity to develop the ... interface between the Meteorological Service and the user of meteorological information. ... Whoever develops this interface may well control the market place for weather information.

3. The United States Approach

In contrast to the situation in New Zealand, the *official* concern of the United States on climate matters is noteworthy. Robert Bergland, former U.S. Secretary of Agriculture, stated in 1977,[30] 'I've issued a ruling that never again do I want to see an economic report come out of the Department of Agriculture that assumes average weather'. The U.S. Interdepartment Committee for Atmospheric Sciences also issued in 1977 a comprehensive United States Climate Program Plan to guide the response to the growing number of climate-induced problems and to anticipate the national and international effects of fluctuations in climate.

The resulting *National Climate Program Act* was signed by the President of the United States in 1978 and in some specific areas progress and developments since then have been impressive. The U.S. Department of Commerce, for example, through its National Oceanic and Atmospheric Administration (NOAA) established a Center for Climatic and Environmental Assessment (CCEA). This Center was an organizational response to two questions: first, how does the United States extend her abilities for observing and predicting short-term weather events to predicting longer-term fluctuations, and second, how does the United States use the knowledge gained in crop production estimates, resources management, energy utilization, etc. The establishment of this new Center permitted a range of new products and services to become available, notably, early warnings and assessments of crop yields, global weather briefings linking climate and socio-economic systems, and assessments of the risk of damage to national resources by climate variations. Regretably this Center is now closed.

4. The Universities

Universities have a very important function in the higher education aspects of meteorology and climatology. The status of meteorology and climatology in many universities however is low, and in some cases non-existent despite the excellent work done by many meteorologists, climatologists, and associated academics in a variety of university departments. For example, in New Zealand, there is no university department of meteorology and/or climatology, and there is considerable reluctance for any of the six universities in New Zealand to give *real* recognition to the role that the meteorological and climatological discipline could play in their educational environment. The disparity between the university situation in a few countries such as the United States where the importance of meteorology and climatology is clearly recognized and appreciated, and that of many other countries such as New Zealand is cause for concern.

An additional factor that should be emphasized is that many national meteorological services continue to appoint professional staff who are exclusively qualified in mathematics and/or physics, and give few opportunities to qualified applicants in engineering, statistics, economics, agriculture, resource management, or geography. Bernard[31] commented in 1976 that it is important for national meteorological services to appoint those who have expertise specially orientated towards the applications of meteorology to development, such as agronomical engineers, hydrologists, and geographers. Indeed, Bernard rightly stated that '...the purely physical and mathematical approach of conventional meteorologists results in their being too impervious to the scientific and technical applications of meteorology for socio-economic progress.' Fortunately, in some countries these suggestions are being acted upon by the universities and the national meteorological services, and the current situation is much improved.

E. NATIONAL SENSITIVITY TO WEATHER AND CLIMATE EVENTS

The economic and social impact of specific national weather or climate events is difficult to quantify. However the direct costs of the 1980 heat wave and drought in the United States are estimated to have been $US11,000 million for major crop losses, $1,000 million for livestock and poultry losses, $1,000 million for increased federal, state, and local government expenditures, $800 million for increased power consumption (mainly additional air conditioning), $600 million for increased requirement for health services, plus a further $3,000 million for other sectors.[32] These dollar losses are very high and represent on a per capita basis nearly $US100 for every American.

The overall sensitivity of nations to weather and climate is much more difficult to assess, especially if all nations are considered. One study, based on the use of each nation's exports as an indicator of weather (and possibly short-term climate) sensitivity has however been made.[33] The analysis considered mid-1970 data from 133 countries and showed that the 'export climate sensitivity values' at that time varied from a maximum possible value of 1,000 in Gambia, to a relatively low value of 198 in Japan. Several major agricultural exporting countries had relatively high rankings out of the 133 countries surveyed (for example, Argentina was ranked 31st, Australia 61st, United States 83rd), whereas several oil exporting countries had relatively low rankings (for example, Venezuela 126th, and Saudi Arabia 130th). There are many overtones to such sensitivities, notably the need and ability for nations to import and export food, but it is considered that the ranking and sensitivity index provides an initial estimate of the weather and climate sensitivity of nations. International agencies may find the rankings useful, particularly if they are involved in climatically related activities such as financial or food assistance programmes.

A comprehensive study of the impact of climate variability on industry in the United Kingdom by the Atmospheric Impacts Research (AIR) Group of the University of Birmingham[34] should also be noted.This 1987/88 survey covers the water supply, construction, energy supply, transport, and the insurance industries, and could well be used as a model by other groups in assessing the impact of climate variability in other countries.

Chapter III

THE GLOBAL SCENE

A. THE INTERNATIONAL SCENE

1. International Responses and Concerns

During 1974 three important world conferences on population, energy, and food took place which realistically considered the changes needed to ensure the optimum use of the world's resources to meet future demands. These conferences were held primarily because of the growing concern about the physical limits to the availability of resources. In 1973 the world received warnings of some of these limits, particularly in the supply of energy and food, and it became apparent that our ability to make efficient and humane use of the world's resources through existing institutions was more limited than we had thought.

In considering the atmosphere as a resource, three factors need to be considered: monitoring and understanding its variability, predicting its variability, and assessing the impact on consumption and production of this variability. But since the impact of short-term variations in the available atmospheric resources can be expected to continue to increase in importance due to the growing demands for food and energy, it is important to understand that *any change* in the climate could well result in much more significant food shortages and energy supply problems than have occurred so far.

Therefore, appropriate international meteorological planning must be evolved if we are to live within our atmospheric income. If this is to be accomplished the international politician and the planner must become more weather orientated, for only then will optimum use be made of the climate resources of the 1990's and beyond. A comprehensive monitoring and analysis of the world's weather and climate to detect and predict changes, and to understand the consequences of such changes, is essential.

2. World Meteorological Organization

The World Meteorological Organization (WMO) is a specialized agency of the United Nations that provides, among other things, an open forum for discussions on a wide range of meteorological and climatological matters. Most importantly it allows the small nations of the world to have a considerable influence on educational, planning, and organizational matters. Particularly in regard to standards of observations and the free exchange of information WMO is well respected, but perhaps the greatest contribution of WMO is through its secretariat, officers, and its technical committees and commissions, who are very aware of the economic, social, and political implications of meteorology and climatology.

The major work of the WMO is carried out by its eight Technical Commissions, each headed by an elected President and Vice-President. One of these Technical Commissions is that relating to climate and the history of this commission provides an interesting international response to the changing nature of climate and climatology. At the first session of the Commission held in 1953 it was called the 'Commission for Climatology', and this name remained for 20 years, but at subsequent meetings held from 1973, the name has been changed to the 'Commission for Special Applications of Meteorology and Climatology', the 'Commission for Climatology and Applications of Meteorology', and finally in 1983 the climatological circle was made complete by the Commission reverting to the original name, the 'Commission for Climatology'.

The new 'Commission for Climatology' is quite different from the earlier Commissions of the same name, primarily for two reasons: first, many climatologists and meteorologists now have a much more enlightened viewpoint of climatology and the climate system; second, the needs of people for weather and climate services and information are now very different from those of the past. The *new* 'Commission for Climatology' can be expected to play an important role in shaping *both* climatological and meteorological decision-making during the coming decades.

3. *Weather and Food: A Case Study Response of International Organizations*

In the early 1970's world agricultural production was depressed as a result of unfavourable weather in several countries which led to a significant decrease in world food reserves, and a corresponding increase in food prices. An important aspect of the global food supply situation is the role of the meteorological community in supplying information which may or may not affect the market. At the international level, the implications of one country being able to monitor and predict and use to advantage another country's weather (and therefore infer that country's potential crop production) are wide-ranging and involve political and in some cases military implications. The international implications of providing and controlling the availability of relevant weather information can be considerable.

In November 1974 at the World Food Conference a number of resolutions were adopted which called upon WMO to take certain actions. Included in the resolutions endorsed by the UN General Assembly was one which requested WMO, in co-operation with FAO, to strengthen the global weather monitoring systems, and to encourage investigations on the assessment of the probability of adverse weather conditions occurring in various agricultural areas. A new and expanded WMO Agrometeorological Programme in Aid of Food Production resulted from these resolutions, the new Programme being submitted to the Seventh WMO Congress, held in Geneva in 1974.

Many of the proposals were supported, but many countries believed the provision of weather-based crop-yield assessments to be 'premature', based on the view that agrometeorological expertise was not sufficiently well developed to provide crop-yield assessments on a regional and global basis with any real degree of precision. But perhaps a more fundamental reason for the non-acceptance of the plan was (and still is) the far-reaching economic, social, political, and even strategic aspects of such information if it is used in the *wrong* way. Indeed weather information can become almost a too powerful tool which can, in the minds of some political leaders, create rather than solve problems.

Considered from an international standpoint, it was clear that there was a need for such information, and the author questioned - at that time - as to how long a system for predicting agricultural production from the already available weather data of the global weather system could remain 'suppressed'. As it happened the whole question of climatic monitoring became more acceptable in the post-1983 period, because countries realized that they could not remain *climatic islands*.

4. World Climate Programme

The World Climate Programme (WCP) was set up following the first World Climate Conference held in Geneva in 1979 in response to the increasing recognition of the need for a more comprehensive monitoring and analysis of the world's climate. It was also partly a reaction to the relatively poor track record of meteorologists engaged in forecasting the weather at various time-scales. In fact, although the aims of the two major global weather programmes, notably the Global Atmospheric Research Programme (GARP) and the World Weather Watch (WWW) were and still are laudable,[1] the benefits have not always been as great as many would have liked. GARP, WWW, and similar *weather* programmes were also basically responses to the concerns of meteorologists and not necessarily responses to the concerns of society.

Central to the aims of the World Climate Programme is the need for comprehensive monitoring and analysis of the world's climate, to detect and predict changes, and also to assess their impact. But while the aims of the World Climate Programme are laudable, the comments made in 1982 by White[2] are noteworthy.

The World Weather Program was just that - a 'weather' program. It was directed at improving our ability to forecast the weather. ... The objective of the program was to improve weather services with very little attention as to how these weather services would be applied to serve society. The World Climate Program however is not just a scientific effort, but also an effort to understand the social and economic implications of climatic variability and change. ...But if the World Climate Program is to be of utility to Governments and others who must make decisions, then it seems to me it must provide information about social and economic consequences. It has to answer the question 'So What?'

The World Climate Programme has four components: data, applications, research, and impact studies. The overall programme is also concerned with applications to agriculture, energy, transport, people, climate prediction at various time scales, and socio-economic studies. The Programme also assists governments in incorporating climatic considerations into the formulation of national policies. In all instances the key factors are relevance to the society we live in and the benefits of the programme to people.

B. SPECTRUM OF WEATHER AND CLIMATE VARIATIONS

1. An Overview

An important consideration of the mix of weather and economics is the ability to use weather/climate information to adjust or modify economic indicators on the basis of their sensitivity to weather and climate. For example, as discussed in Chapter X the *Business Week Index* of United States economic activity can be adjusted in such a manner that economic activities such as electricity consumption more truly reflect the important environmental conditions rather than just pure economic 'strength'. This concept of 'weather adjusting' economic indicators is often considered to be beyond the scope of both the meteorologist and the economist. Nevertheless, weather information suitably weighted and specifically adjusted to take into account various economic distributions such as population, energy consumption, corn production, electricity production, etc., is available for such adjustments, and these adjustments are essential if a proper interpretation of weather/climate sensitivity is to be assessed.

An additional refinement of weather/climate adjusting is the use of commodity-weighted weather information in *forecasting* the trend of national economic indicators. Weather information is available in real-time, whereas national economic indicators usually have a time publication delay of several weeks before the actual production/consumption information is available. Attempts to make weather/climate adjustments to national economic indicators and to make weather-based forecasts of such indicators are difficult and controversial. However, it is considered that they not only provide economists, decision-makers, and publishers of economic indicators with realistic adjustment factors, but also provide valuable assistance for the economic forecaster.

2. The Atmospheric Component

Extreme weather events - hurricanes, tornadoes, lightning, floods, droughts, and hail - often result in catastrophic losses of property and income, and in loss of life. In the United States, for example, the cost of hurricanes in an extreme year is likely to be in the $5,000 million to $10,000 million range. Nevertheless the much more usual 'above average' and 'below average' weather and climate conditions are also economically significant, and it is suggested that the weather and climate events which *do not* make newspaper headlines are more important in the long run than those that do.

A significant question therefore is whether it is the relatively small changes in the overall climate, rather than the extreme events (or alternatively, whether it is the 'small' long-term events, or the 'larger' short-term events), which have the more important overall social, economic, and political impact. For example, a drop of 3°C in summer temperatures in the Soviet Union, a 10% decrease in cloud amount over North America in the winter, or a 1°C increase in the summer temperature of the South Pacific Ocean (if sustained for more several seasons) could well have social, economic, and political consequences which in total could be much greater than the destructiveness of hurricanes, severe rain storms, or tornadoes.

3. Political and Marketing Realities

In the past many meteorologists and climatologists were quite content to remain ignorant of the economic, social, and political value of their work, but it is now evident that meteorology and climatology must provide appropriate returns - social, economic, and political. But while budgetary considerations are important, the more fundamental reason for being interested in the impact of weather and climate information is that nations will undoubtedly try to apply the flow of weather information in a more deliberate manner, in order to control significant weather-sensitive aspects of other national economies.

The deliberate modification of the weather and climate of one country by another country may also become a reality, and although such modification will - at least in the foreseeable future - be limited in area, its effect on certain aspects of agriculture could be considerable. Monitoring and control of the atmospheric environment are therefore matters which will continue to have increasing economic, social, and political consequences at the highest level.

It is relevant to note in this regard that few if any of the models of the physical atmosphere go on to consider economic and social aspects. Until the 1970's only token consideration was given to the end products of the meteorological chain, but it is now recognized that weather and climate information has social, economic, strategic, and political value, and that such information must be correctly presented to the decision-maker. The meteorological and climatological community therefore has a continuing task of providing ways to establish direct communication between top-level decision-makers and the meteorological system.

C. GREENHOUSE GASES / CLIMATE CHANGE DILEMMA

1. An Overview

During the last decade there has been increasing concern - at least among the scientific community - over the impact of carbon dioxide and other greenhouse gases on global atmospheric temperatures and sea level. This concern is now reaching the public at large, and in particular politicians and planners are asking for guidance as recently demonstrated in a *Time* (19 October 1987) cover article entitled 'The heat is on '. Among other things *Time* stated:

> Until now, the earth's climate has been a remarkably stable, self-correcting machine, letting in just the right amount and type of solar energy and providing just the right balance of temperature and moisture to sustain life. Alternating cycles of cold and warmth, as well as greater and lesser concentrations of different gases, have forced some species into extinction. The same changes have helped others evolve. The irony is that just as we have begun to decipher the climatic rhythms that have gone on for hundreds of millions of years, we may have begun to change them irrevocably... .

At a joint UNEP/WMO/ICSU conference on 'An assessment of the role of carbon dioxide and of other greenhouse gases in climate variations and associated impacts' held in Villach (Austria), from 9-15 October 1985, scientists from 29 developed and developing countries assessed the role of increased carbon dioxide (CO_2) and other radiatively active constituents of the atmosphere (collectively known as greenhouse gases) on climate changes and associated impacts. They concluded[3] that as a result of the increasing concentrations of greenhouse gases it is now believed that in the first half of the next century a rise of global mean temperature could occur which is greater than any in man's history. However, the subject is not without controversy, and it is perhaps relevant to give a brief insight into some other viewpoints.

In a research review in 1984 on 'The CO_2 climate controversy: An issue of global concern', Idso[4] made the following comments:

> The CO_2 greenhouse hypothesis was first put forward by J. Tyndal[5] in 1861, and since that time it has provided a provocative stimulus for many students of the atmosphere. Although a wide range of investigators subsequently produced a wide range of estimates relative to the magnitude of the phenomenon, a consensus was finally forged by a group of climate modellers meeting under the auspices of the U.S. National Research Council in 1979.[6] Based on computer

calculations of the general circulation of the atmosphere, this group concluded that a 300 to 600 ppm doubling of the atmospheric CO_2 concentration would lead to a 3 +/- 1.5°C increase in mean global air temperature.

Idso then discussed in detail many aspects of the CO_2 issue, and commented that there is no *a priori* reason to believe that the computer models of the atmosphere currently employed to investigate the effects of increasing atmospheric CO_2 are reasonable representations of reality. Indeed, as he says: '... by the admissions of their own creators, they fall far short of that ideal'.

In contrast, Tucker,[7] in discussing 'The global CO_2 problem' in 1985 with particular reference to Australia, took a more 'middle-of-the-road' viewpoint. He comments on the work of Monteith,[8] on the sensitivity of crops to climatic variation, and quoting Monteith says that: 'Until the predictions from climatic models become more reliable, I see little point in developing "scenarios" for agricultural production based on numerous insecure premises.' Tucker points out however that the range of plausible future CO_2 - climate scenarios is now at the stage where more detailed response assessments could be commenced, but he cautioned that care must be taken to sift carefully diverse estimates of the future climate, and to consider as the basis for consequential studies only those that are founded on sound experiment and reasoning.

2. 1985 Villach Conference Findings

Among the principal findings of this Conference previously referred to were:

Many important economic and social decisions are being made today on long-term projects such as irrigation and hydro-power, drought relief, agricultural land use, structural designs and coastal engineering projects, and energy planning - all based on the assumption that past climatic data are a reliable guide to the future. This is no longer a good assumption since the increasing concentrations of greenhouse gases are expected to cause a significant warming of the global climate in the next century.

The role of greenhouse gases other than CO_2 in changing the climate is already about as important as that of CO_2. If present trends continue, the combined concentrations of all of the greenhouse gases would be equivalent to a doubling of the CO_2 concentrations from pre-industrial levels possibly as early as the 2030s.

While other factors such as aerosol concentrations, changes in solar energy input, and changes in vegetation may also influence climate, the greenhouse gases are likely to be the most important cause of climatic change over the next century.

There is little doubt that future changes in climate of the order of magnitude obtained from climate models for a doubling of the atmospheric CO_2 concentration could have profound effects on global ecosystems, agriculture, water resources and sea ice.

Governments and regional and inter-governmental organisations should take the results of the Villach 1985 assessment into account in their policies for social and economic development, environmental programmes and policies for the control of greenhouse gas emissions.

Work should be started immediately on the analysis of policy and economic options, when the widest possible range of social responses aimed at preventing or adapting to climate change should be identified, analysed and evaluated.

3. 1987 Villach Workshop on Climate Change

The scientific consensus discussed above was used as the starting point for the Villach Technical Workshop on 'Developing Policies for Responding to Climatic Change' held in September 1987[9] with the following main themes: possible scenarios for future changes of climate and sea-level; effects of possible climate changes on regions in the high latitudes, middle latitudes, humid tropics, semi-arid tropics, and coastal zones; management options for responding to the possible changes; and considerations that might bear on policy development.
    The importance of considering the rates of change of climate, sea-level, sea-ice, vegetation, etc., was also considered, and the Workshop found that the concept of setting limits on the rates of change might have important implications for the choice of policy options. The Workshop also gave a clear message that scientific assessments of climate change must address three issues if they are to be useful for policy discussions: the rate and timing of climate change; the uncertainties in forecasts of climate change; and the regional distribution of climate changes.
    The present scientific understanding of the impact of the greenhouse effect on global temperatures as reported in the findings of the 1987 Villach Workshop is shown in Fig. III.1. This shows a range of scenarios of global temperature change that might plausibly occur between now and the end of the next century. The historical record of actual temperature changes that have occurred since 1850 is included for perspective. All scenarios quickly carry the world into a condition of of higher temperature and sea level than has been experienced within the last 100 years. The intensity of the global hydological cycle in terms of precipitation and evaporation is expected to increase by 2-3% for each degree of global warming. The changed global climate can therefore be expected to be wetter than that of the recent past.
    Allowance has been made in Fig. III.1 for emissions of all the significant greenhouse gases affecting climate (including the chlorofluorocarbons) and for

Figure III.1 Global temperature change (°C)

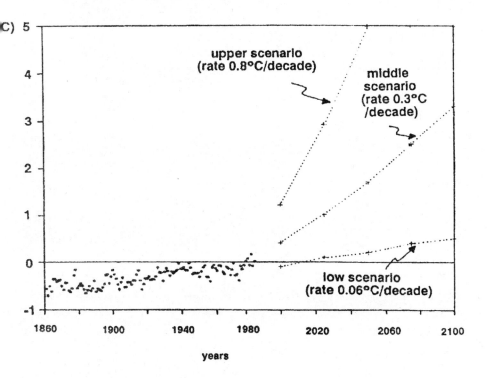

Source: World Meteorological Organization (1988)

time-lags in the climate's response introduced by the ocean's heat capacity. The climatically induced sea-level changes shown in Fig. III.2 would be the same everywhere on earth, although they may also be affected by local geological (tectonic) phenomena. The temperature changes, in contrast, reflect the annual average conditions of the world as a whole, not the conditions at any specific place.

The wide range of scenarios depicted in the diagrams reflect uncertainties in two basic areas: future patterns of fossil fuel use, rates of reforestation and de-forestation, and other activities leading to greenhouse gas emissions; and the response of the climate system to a given level of greenhouse gases. These uncertainties contribute about equally to the overall uncertainty in forecasting future climate change. The envelope of scenarios pictured in Fig. III.1 and Fig. III.2 has been constructed so that, in the judgement of the 1987 Villach Workshop, there is a 90% chance that the actual future pattern of climate change will lie within the bounds set by the upper and lower curves.

The upper curve represents the temperature and sea-level change that could result if there is a radical expansion of fossil fuel use and other activities that emit

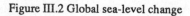

Figure III.2 Global sea-level change

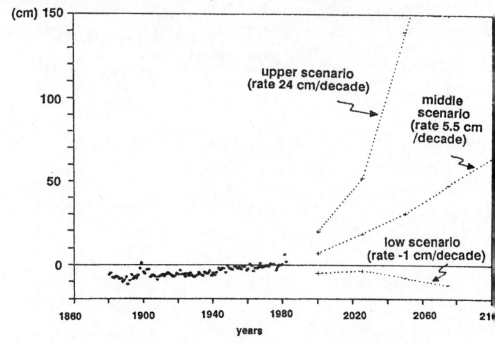

Source: World Meteorological Organization (1988)

greenhouse gases, and if the climate's response to greenhouse gases exhibits the high sensitivity predicted by a few studies. In contrast, the lower curve represents the change that could result if there is a radical curtailment of fossil fuel use and other activities that emit greenhouse gases, and if the climate's response exhibits the relatively low sensitivity predicted by a few other studies. The middle curve shows the temperature and sea-level change that could result if there is no change in the present rates of increase of greenhouse gas emissions, and if the climate response exhibits the moderate sensitivity to greenhouse gases that the majority of studies predict. In calculating this curve the Villach Workshop assumed that the 1987 Montreal protocol on protecting the ozone layer will be successfully implemented, thus reducing the emissions of chlorofluorocarbons significantly below their recent rates of increase.

All scientific studies of the greenhouse effect indicate that the resulting climate changes will vary among regions (the one exception being the climate-induced component of sea-level rise which should be the same for all stable coastal areas). Uncertainties in the forecasts of regional climatic responses are even greater than those in the forecasts of global climate response, and the present thinking about possible regional climate changes is summarized in Table III.1. This indicates that

Table III.1 Regional scenarios for climate change

Region	Temperature change *		Precipitation change
	Summer	Winter	
High latitudes (60 - 90°)	0.5x - 0.7x	2.0x - 2.4x	Enhanced in winter
Middle latitudes (30 - 60°)	0.8x - 1.0x	1.2x - 1.4x	Possibly reduced in summer
Low latitudes (0 - 30°)	0.7x - 0.9x	0.7x - 0.9x	Enhanced in places with heavy rainfall today

Source: World Meteorological Organization (1988).
Note: * as a multiple of the global average (x).

the greatest warming is likely to occur during the winter in the high latitudes of the Northern Hemisphere, with changes 2.0 to 2.4 times greater and faster than the globally averaged annual values shown in Fig. III.1. In contrast, temperature changes in the low latitudes will probably be somewhat smaller and slower than the globally averaged values. Regional precipitation forecasts are the most uncertain of all. However, the studies suggest that major changes could include enhanced winter snowfall in the high latitudes, intensified rains in the presently rainy low latitudes and, perhaps, a decrease in summer rainfall in the middle latitudes.

4. Management Options for Responding to Climate Change

Strategies for responding to a changing climate fall into two categories. Limitation strategies slow or reverse the growth of greenhouse gas concentrations in the atmosphere, whereas adaptation strategies adjust the physical environment or our ways of using it to reduce the consequences of a changing climate. A prudent response should consider both approaches.

Adaptation will involve expenditure of very large financial resources over long periods of time with extended advance planning. Measures to adapt to climate change will occur on a variety of scales and with widely varying costs. Some environmental modification measures, such as changes in coastal defences and freshwater supply systems, require infrastructure investments beginning decades in advance of anticipated climate impacts. On the other hand, many adjustments will consist of behavioural changes at the individual level occurring in immediate response to perceived warming, with little advanced planning.

Various limitation options exist for achieving CO_2 reductions, in spite of continued population growth, and in a manner consistent with continued economic expansion. These include a reversal of the current deforestation trend, efficiency advances in generation and transmission of energy, shifting the fossil fuel use mix from high to low CO_2-emitting fuels, disposal of CO_2 in the deep ocean, and replacing fossil fuel combustion with alternative energy sources such as solar, wind, hydro-electric, nuclear, tidal, and ocean thermal conversion.

However, even if limitation and adaptation strategies are found to give cost-effective management solutions to the climate change problem on a global scale, the actions to achieve both strategies will have to be taken at national and local levels and in many cases in accordance with internationally agreed guidelines. Moreover, every nation will differ in the economic and social costs which it has to devote to limitation and adaptation strategies. These costs will depend on a number of factors: the degree and type of industrialization and fossil fuel use, and the gross domestic product per capita; the type and level of nitrogenous fertilizer used in agriculture; the geographical location of the country; and whether there is a coastline (including how much of it is low lying, built over, or used for agriculture or tourism).

Some countries believe that they will be 'winners' and some believe they will be 'losers' with a problem not of their own making. At present, the information available on climatic impacts is not sufficient to suggest clear winners and losers, but some countries could well perceive that they will be.

5. Timing of Responses to Climate Change

The best estimates of global mean temperatures over the last 100 years suggest that overall they have risen about 0.5°C, and that sea level has risen during the same period by about 12 cm. It is further considered that the current atmospheric levels of greenhouse gas emissions may already have committed the world to an additional 0.5°C of warming, and an additional 10-30 cm of sea-level rise over the next fifty years, *even if the atmospheric composition were stabilized immediately.*

An additional factor of considerable concern is that our understanding of the atmosphere and the oceans and natural ecosystems is limited, and the possibility exists that large and abrupt changes may occur which could overwhelm our adaptive capabilities, which could far exceed the natural rates of change in ecosystems. The likelihood of such a 'climatic surprise' increases as climate deviates from historical bounds.

The opinion of most scientists is that it would be inappropriate to postpone action on the greenhouse gas/climate question until the consequences of warming, which lag behind greenhouse gas emissions, are clearly visible. But policy actions already implemented, or under consideration, could limit or mitigate the consequences of warming. The timing of these initiatives to limit and adapt to climate change is critical from both an environmental and a financial perspective.

6. Costs of Responding to Climate Change

If current perceptions about the range of possible environmental and socio-economic effects are substantially correct, then the costs incurred by doing *nothing* about climatic change could be extremely large. However, strategies for either adapting to climatic change or limiting it by controlling greenhouse gas emissions, or both, will also involve high costs to global society. Clearly, it would be preferable to be more certain about the magnitude and rate of onset of global warming and about its environmental and socio-economic effects before taking expensive adaptation and/or limitation actions. But time is not on our side, and at least for policy-making purposes there is an urgent need for detailed comparisons of the costs of various strategies.

Some of the items that need to be included in any programme for comparing the costs of different strategies are shown in Table III.2. This considers three scenarios: 'Business as Usual', 'Moderate Effort', and 'Concerted Effort'. The different scenarios reflect the level of effort and ambition of policies specifically undertaken to deal with climate change induced by greenhouse gas. The 'Business as Usual' scenario implies that no policies explicitly directed at greenhouse gas limitation are undertaken, whereas 'Moderate Efforts' and 'Concerted Efforts' reflect the level of effort devoted to such items as energy policy, reforestation, and greenhouse gas reduction strategies. The 'Surprise' scenario differs from the other three scenarios, since it could occur in any one of the other scenarios, although it is perhaps less likely to occur in the case of 'concerted efforts' than that of 'business as usual'. It is intended to highlight the consequences of an unpredicted surprise event, such as an abrupt change of climate as a result of an unpredicted change of the ocean circulation.

The response 'Limitation' refers to specific adaptation steps taken *before* effects occur, while forced adaptation occurs in *response* to physical and biological impacts. Residual costs are those for which adaptation steps are not, or cannot be, undertaken, and will largely be 'external' costs involving such things as the global commons, unmanaged ecosystems and human suffering. It should be emphasized that ultimately adaptation strategies will have to be replaced by limitation strategies, since a significant warming would sooner or later become intolerable no matter how much is spent on adaptation. Limitation strategies on the other hand cannot totally limit emissions and some adaptation will be required, especially as a result of the emissions that have already occurred.

7. The Role of the Market and International Initiatives

A preference for adaptation as a strategy in the greenhouse gas/climate change dilemma is often expressed by those who consider that free market forces will take care of the problems of adaptation, as and when they may occur, whereas this could not happen with a pre-planned limitation strategy. The difficulty here is that the

Table III.2 Relative costs* of four different types of effort undertaken in three different strategies for responding to climate change

Strategy	Limitation costs	Anticipatory adaptation costs	Forced adaptation costs	Residual costs
	(long lead time)	*(varying lead time)*	*(no lead time)*	
Busines as usual	w	xx	yyyyy	zzzzzz
Moderate efforts	ww	xxxx	yyy	?
Concerted efforts	wwww	xx	y	z
Surprise			yyyyyy	zzzzzzzz

Source: World Meteorological Organization (1988).
Note: * The relative costs are indicated by w, x, y, z. The costs are given only to indicate relative costs within each column. The costs in the limitation column would be investment costs that could become net positive investments (that is generators of profit), whereas costs for adaption would in many cases involve repairs to damage.

long time-frames of climate and ecological processes do not fit into the much shorter time-frames of economics where events 10 years off can have their values discounted by 65% and those 40 years off by 98%. It is considered that the discounting problem will ensure that the role of free market mechanisms will at best be marginal, and cannot take the place of government actions at both national and international levels.

Discussions on strategic policy issues in connection with the greenhouse gas/ climate change question is still at a very basic stage. The Villach 1987 Workshop saw itself as trying to take a first step in this direction, and recognized that all the ideas discussed needed a long period of much more careful development before sensible and robust policies can be identified and agreed upon internationally. An initial follow-up activity took place in Bellagio (Italy) in November 1987 at which 25 experts discussed the formulation of policies for responding to future climatic changes. The findings of this meeting[10] were first used as an input to the December 1987 meeting of the International Advisory Group on Greenhouse Gases (AGGG) of WMO, UNEP, and ICSU. It was also an important input to a Canadian sponsored conference on 'The Changing Atmosphere, Implications for Global Security'[11] held in Toronto in June 1988, and the Commonwealth Expert Group on Climatic Change and Sea Level.[12] These activities will in turn provide input to the Second World Climate Conference of the World Meteorological Organization expected to be held in 1990.

The problem of significant climate warming calls for a considerable increase in global monitoring activities, and the further development of climate models to improve our understanding of, and reduce our uncertainties about, the extent of regional and global climatic change and its impacts on the environment and major socio-economic sectors. In particular, the global scientific community needs to carry out studies to answer three questions: what new climate observing system activities are required for monitoring the changing global climate; what basic emissions data on greenhouse gases need to be continuously monitored and archived globally; and what activities are required for monitoring the consequences of the changing climate?

Clearly the greenhouse gas/climate change dilemma will be a very significant research and impact problem in the years ahead, and the many investigators in the area will need to provide guidance to decision-makers in a very positive manner if, as it is predicted, a significant warming does occur.

D. OZONE - ENVIRONMENTAL AND CLIMATIC EFFECTS

1. Ozone: Creation and Destruction

In its natural state atmospheric ozone (0_3) occurs in a layer which has a maximum concentration at about 25 km above the surface of the earth. It is constantly being created and destroyed through natural chemical cycles. The layer is critical to life on earth because it filters out damaging ultraviolet radiation from the sun.

During the 1920's, chlorofluorocarbons (CFC's) were invented, and until the 1970's they were considered to be the ideal chemical for numerous industrial and consumer applications.   Being inert, non-toxic, and cheap, they became almost indispensable for refrigeration, foam blowing, aerosol propellants, fire extinguishers, and as solvents. However, a possible problem with CFC's was recognized in the early 1970's, when it was realized that chlorine compounds (such as CFC's) produced by industry could deplete the ozone layer in the stratosphere.

Specifically, when a CFC molecule is released and eventually carried up into the stratosphere, it is decomposed by solar ultraviolet radiation and produces an 'unattached' chlorine atom. This atom can initiate an ozone-destroying reaction sequence while re-emerging unchanged itself.   The single chlorine atom can emerge from the reaction sequence perhaps 100,000 times before it is finally removed by something else. Accordingly, the 100 grams of CFC in a single spray can, or a single domestic refrigerator, can eventually destroy over 3 tonnes of ozone.[13] The milions of tonnes of CFC's currently in the atmosphere will continue to 'leak' into the stratosphere and affect the ozone for the next 200 years. Already, observational data from many parts of the earth are showing a small downward trend in ozone.

2. The Antarctic Ozone Hole

In 1985, a striking and alarming phenomenon was first recognized[14] which has become known as the 'hole in the ozone layer'. The 'hole' occurs over the Antarctic continent in the spring, during which ozone is depleted for about two months. The feature seems to have occurred first in the late 1970's and there has been a definite trend towards a bigger 'hole' since then. The 1987 'hole', which persisted longer than in earlier years, had less than half the 'normal' amount of ozone, and (as of September 1988) was the most severe 'hole' so far observed.[15] It is now widely accepted by scientists that the 'hole' is due to chlorine from CFC's, together with the special meteorological conditions associated with Antarctica.

3. The Montreal Protocol[16]

An agreement, called the *Montreal Protocol on Substances which Deplete the Ozone Layer*, was signed, subject to ratification, by 24 countries on 16 September 1987. The Protocol is a complex legal document but the main points are: (1) by 1990, all Parties to the Protocol will freeze CFC consumption to the same levels as 1986, (2) by 1994, all Parties will reduce CFC consumption to 80% of the 1986 levels; and (3) by 1999, all Parties will reduce consumption to 50% of the 1986 level.

If these consumption levels are achieved, it will mean that chlorine amounts will by 1999 increase by only 2% per year, compared with an increase of 5% per year if there was no Protocol. The Protocol will therefore limit the worsening of the problem. Clearly damage to the ozone layer will still continue, but there is now a clear signal to industry that action to develop alternatives and new technology must start now. Moreover the world community has shown a definite will to prevent continuing atmospheric environmental damage.

There are various provisions for continuing scientific assessments and reviews of the Protocol timetable. It is important to note in this regard that the Protocol is *consumption* based rather than *production* based. This means that the burden of compliance is shared more evenly between the producing nations and those which import all their CFC's.

4. Ozone Impacts

The main long-term effect of reduced ozone amounts is an increase in ultraviolet radiation near the earth's surface. This can lead to an increase in skin cancer, the degradation of many materials such as plastics, paints, and fabrics, and the reduction in productivity of significant crops such as rice, and in oceanic plankton concentrations.[17] In addition, it is important to note that while the concentration of carbon dioxide is increasing at about 0.5% per year, the concentrations of *other* greenhouse gases are also increasing as a result of human activities. In particular methane is increasing at about 2% per year and has doubled in the last 100 years, and as noted above there has been a dramatic increase in CFC's. These and other greenhouse gases are currently at lower concentrations than carbon dioxide, but they are increasing at such a rate that their contribution to the greenhouse effect will be equal to that of carbon dioxide within 50 years.[18] Clearly, control of CFC's will assist in reducing the *overall* greenhouse effect discussed in detail in the previous section.[19]

E. WEATHER AND CLIMATE MODIFICATION

1. An Overview

Weather and climate affect human activities in pervasive ways. For example, the profitability of many economic activities depends very much on the weather and climate, obvious examples being agriculture and air transportation. Some human behaviour patterns are also influenced by weather variations. It is natural therefore that people should try to find ways of not only adjusting to the weather and climate, but also altering the atmospheric environment in which they live.

Two major types of adjustment action can be taken: first, the adoption of techniques which 'insulate' activities from weather and climate variations, such as using air conditioning or growing drought-resistant crops; second, alterations in patterns of activity may be made, such as the modification of a harvesting schedule, the postponement of a shopping trip, or temporary movements away from an area, such as would occur in the case of evacuation from the path of a hurricane. Alternatively, or in addition, people may deliberately or inadvertently try to alter the amount or distribution of particular weather elements such as precipitation, storminess, high temperatures, or frost.

People and nations have generally concentrated their attention on the first alternative; that is, moving to those locations where the weather and the climate is both conducive to their personal comfort and where economic activities can be pursued for maximum benefit, as well as trying to develop ways of reducing the impact of weather and climate variations if it is not possible for them to move. However, there is now acceptance that in some areas, under certain circumstances, and at specific times, weather modification in the form of an increased precipitation of up to 10% can be achieved, and that the dispersal of some kinds of fog is feasible. However, there is still considerable uncertainty as to how 'successful' weather modification has been, particularly if *all* of the impacts on an area are considered.

2. *Weather Modification and Tomorrow's Weather: Essential Linkages*

Modification of the atmosphere - whether intentional or otherwise - can be done in a few areas and during certain time periods. Little thought, however, has gone into some of the fundamental questions regarding such modification. For example, while the basic scientific question may well be 'can you modify the weather', a more fundamental question is 'should you do so', 'where should you do so', and 'what safeguards are there either in the form of laws or compensation payments to provide for the "errors" which may occur'? Some of these aspects were first explored in a symposium held in July 1965, which culminated in the publication of a monograph *Human Dimensions of Weather Modification;*[20]later the National Center for Atmospheric Research established a *Task Group on the Human Dimensions of the Atmosphere*[21] to explore these matters in greater detail.

Many people consider that weather modification and weather forecasting are very far apart. However modification of tomorrow's weather may be very dependent on the weather that is forecast for tomorrow, since decisions as to whether to modify should be based on what is predicted. It is also important to emphasize that, to the client, a correct weather forecast and successful weather modification may well have similar effects, since the client is not interested in why it rains, but only that it does or does not rain.

3. *The Acid Rain Problem*

'Of all the issues involved in its relationships with the U.S., none - not even free trade - concerns Canada more than acid rain'. So stated *Time* (11 April 1988) in a comment on 'talking tough' on pollution. In a similar vein in a special issue of *Weatherwise*, Miller[22] says that acid rain is a key environmental issue of our time. He points out that this is not just a local problem affecting a few people in isolated industrial areas, but rather a phenomenon of national and international scope that involves very large areas of North America and Europe. He continues:

> By some accounts these affected regions have grown in size from year to year, and the acidity of the rain in some of them may have been increasing as well. The economic and environmental implications of these increases have propelled public debate on the issue to a fever pitch. Politicians have been inundated with appeals for legislative action to curb the problem.

There are three important 'impact' aspects of acid rain: the general degradation of sensitive lakes and rivers, the damage to forests, and the surface 'erosion' of buildings. The ultimate causes of these impacts on the environment are subject to some debate, and further research is needed. But, whatever the cause, the impact of acid rain in the form of 'serious' damage to forests is real and has been to the forefront of several intergovernmental 'exchanges'; indeed it is not unusual for the 'acid rain problem' to be the 'front page' news in national newspapers and on network television in both North America and Europe.[23] Several countries have positive programmes on acid rain, notably Switzerland which is clearly vulnerable to acid rain destruction of her alpine forests. Public information brochures on the problem have been issued including one entitled 'Ses dernieres pousses nous lancent un SOS' published by the Swiss Federal Department of the Interior.

4. Implications of a 'Nuclear Winter'

The impact of a nuclear war on the climate of the earth has been the subject of many reports and research papers since the early 1980's. In a summary of the situation as of 1984, Elsom[24] stated that prior to 1982 climatologists believed that even a large-scale nuclear war involving tens of thousands of nuclear detonations would have little effect on the global climate. However, he noted that such a view was no longer held to be valid by 1984 and he cited several investigations which indicated that a large-scale nuclear exchange may well cause profound changes to the climate of both hemispheres.

Another study by Cess[25] further summarized the position as of the mid-1980's; he noted that following a nuclear exchange there might be a significant reduction in surface temperature over land areas, because of the impact on the radiation budget of smoke produced by fires, and dust injected into the stratosphere. His concluding remarks are significant: 'It is important to re-emphasize that the type of climate response associated with existing nuclear-war climate studies is unique with respect to our experience with climate models .... . This raises the question as to the validity of existing climate models ... .'

Mention should also be made of the report prepared by the Royal Society of New Zealand.[26] This 1985 report describes the threat of nuclear war, possible ways of relieving the threat, and the role of scientists. One chapter of this publication on the climatic effects on the Southern Hemisphere discusses the implications to New Zealand of nuclear explosion in various parts of the world and it states:

... the long term effects on New Zealand of a northern hemisphere nuclear war with some smaller extension south of the equator cannot be given with any certainty but are likely to be much greater than had been thought in recent years. The greatest environmental hazard in the northern hemisphere from a nuclear war is thought to come from the reduction in sunlight produced by smoke clouds. There is a possibility that the southern hemisphere may be exposed to this 'nuclear winter' effect though at a reduced level.

F. CLIMATE CHANGE AND POLITICAL REALITIES

In discussing 'climate change and political realities', many questions are raised; these include what is meant by 'climate'. The current thinking by many climatologists is that 'climate data' include 'atmospheric' data for *all* time scales (that is current, near real-time, and traditional 'historical' climate time scales). This implies that climate is not restricted by any time scale such as the conventional but surely now outdated 30-year 'normal' period, and indeed must now include climate forecasts.

A specific aspect of climate change and political reality is the agricultural consequences of a temperature change. Indeed, the impacts of a significant climate change (through one or several causes including but not restricted to the increasing amount of 'greenhouse' gases in the atmosphere) on agricultural crops in various parts of the world, have been assessed by a number of investigators. One such study by the United States National Defense University[27] showed that a 'large' warming could increase spring wheat yield in Canada by about 8%, but decrease winter wheat yield in Australia by about 4%.

Should such a climate change take place there could be far-reaching economic, social, political, and strategic consequences. Among other things these would involve famine relief, possible large-scale climate modification, international aid programmes, and migration of island groups affected by sea-level rises. Some of these expected consequences have been studied by the International Institute of Applied Systems Analysis,[28] the United Nations Environment Programme,[29] the World Climate Impact Programme,[30] and the Commonwealth Secretariat[31] in London, and these findings are of considerable importance to many countries. However, whatever the impacts of any long-term trend such as a significant warming over say a 50-year period, it is evident that in many cases much more important year to year variations will still occur.[32] Indeed, it is of considerable economic and political significance to emphasize that in most cases, trends in the climate over time which *are* able to be adjusted to, are usually small by comparison with short-term variations. Moreover, even if there is *no* significant climate change, the impact of short-term variations in the available atmospheric resources must continue to increase in importance because of the growing vulnerability of many societies to human and natural environmental changes.

Nevertheless, the 'fear' of a 'climate change' is clearly of great concern to many people - irrespective of the impact of today's weather. This concern reached the front pages of many of the world's newspapers in 1988; indeed the *New York Times* in a front page article on 24 June 1988 stated in a headline:'GLOBAL WARMING HAS BEGUN, EXPERT TELLS SENATE'. The newspaper went on to explain that until now scientists have been cautious about attributing rising global temperatures of recent years to the predicted global warming caused by pollutants in the atmosphere, known as the greenhouse effect. 'But today Dr James E. Hansen of the National Aeronautics and Space Administration told a Congressional committee that it was 99 percent certain that the warming trend was not a natural variation but was caused by a build up of carbon dioxide and other artificial gases in the atmosphere. ' Dr Hansen added : 'It is time to stop waffling so much and say that the evidence is pretty strong that the greenhouse effect is here.'

While it is clear that not all climatologists would agree with Dr Hansen, it is very evident that people and governments around the world want to know more about the potential impacts of any climate change. *Newsweek*, for example, in presumably interpreting the 'feel of the people' carried a special cover story in its 11 July 1988 issue on 'Inside the Greenhouse'. Similarly, *Time* in a cover story on 'The Big Dry' in its issue of 4 July 1988 in commenting on Dr Hansen's testimony to the U.S. congressional committee (as noted above in the report from the *New York Times*) stated: 'Even Hansen's scientific critics hope his testimony, however premature, will prod people into taking measures to ease the greenhouse effect by conserving energy and cutting back on burning fossil fuels.' But *Time* adds : 'The alternative, though, may be even less pleasant for many. As Democrat Senator Wendell Ford of Kentucky pointed out last week, the only major energy source that might replace fossil-fuel plants is nuclear power.'

In any discussion of the 'economic climate', it is essential therefore to understand the increasing impact climate variations will have on economic, social, political, and strategic activities. One positive step in this direction were the findings of the conference on 'The Changing Atmosphere' held in Toronto during 27-30 June 1988. In the Foreword to the Conference Statement , the conference director (H.L. Ferguson) states: ' The message from the Toronto Conference was clear. The Earth's atmosphere is being changed at an unprecedented rate, primarily by humanity's ever-expanding energy consumption, and these changes represent a major threat to global health and security.'

Since most political regimes - both developing and developed - have a 'thinking lead-time' of years rather than decades, any real economic, agricultural, political, or strategic significance of a long-term climate trend is of course very difficult to infiltrate into the political decision-making process. However there are positive signs that attitudes are changing as demonstrated by the address by His Excellency Mr Maumoon Abdgul Gayoom, President of the Republic of Maldives, to the United Nations General Assembly on 19 October 1987. The President stated: 'We in the Maldives have seen and lived through grim experiences which could be the indicators of dire consequences of global environmental change provoked and aggravated by man. ... It is in the interest of all the world that climatic changes are understood and the risks of irreversible damage to natural systems, and the threats to the very survival of man, be evaluated and allayed with the greatest urgency.'

Chapter IV

INFORMATION, COMMODITIES, AND COMMUNICATIONS

A. DATA AND INFORMATION

The preface to *Quarterly Predictions* published by the N.Z. Institute of Economic Research states that '... economic forecasting is a chancy business for in addition to our own imperfect understanding of the way the economy works, there are also likely to be errors from chance factors such as changes in the weather at home and overseas, from inadequacies in our basic information, and from the unknown effects of political events'. In the light of these comments, there are clear opportunities for meteorologists and climatologists to provide decision-makers with more relevant weather and climate information.

To assess the increased economic, political, and social benefit that may arise from an improvement in the use of weather and climate information is however difficult. One approach is to consider that the economic outcome of a weather/climate sensitive process under management is influenced by four factors (Fig. IV.1): actual weather events (that is, what actually occurs); weather information (that is, what is reported to have occurred, or what is forecast to occur); non-weather events (such as actual prices); and non-weather information (such as future exchange rates). In all cases, the economic, social, and political outcome of a weather or climate sensitive process under management is subject to uncertainty, principally because at the time the most appropriate alternative is chosen, the decision-maker does not know the actual value of either the 'weather information box' or the 'non-weather information box'.

Weather and climate information constitute a number of separate items including the weather existing at the present moment, the weather and the climate that is expected at a specified time in the future, and an analysis and interpretation of the records of the weather and the climate that occurred in the past. All three types are of value[1] but the specific information required will be determined by the kind of problem and the associated decision-making involved. For example, an analysis of the past weather and climate is useful for assisting in planning the location and design of irrigation schemes and agri-business enterprises, whereas relevant real-time or near real-time weather information is important for operational type decisions such as marketing, agricultural production forecasts, and analysis of agri-business trends. Similarly, weather and climate forecasts are particularly useful in making operational decisions such as the scheduling of irrigation, estimating energy demand, or forecasting the 'futures' prices of agricultural products.

Figure IV.1 The relationship between weather events and information, non-weather events and information, the choice of alternatives, and the economic outcome of a weather-sensitive process under management

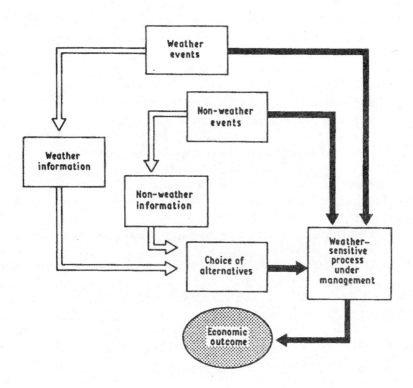

Source: After Maunder (1985), from Maunder (1970) and McQuigg and Thompson (1966).

Applications of meteorological and climatological information have been considered by a number of investigators including those at a special symposium on *The Business of Weather* held over 20 years ago in Chicago. Hallanger[2] made some very pertinent remarks to businessmen at that time which are possibly even more relevant today than they were in 1963. He stated:

> ... the key point in the realisation of the potential value of weather information to your activity (is that) ... the meteorologist, working as a team with your people, must become familiar with your operation. Only then is he in a position to identify the true weather problems. Only then can he provide the appropriate weather information in the most useful form. Only then will the return exceed the cost by the greatest amount.

B. CLIMATE EFFECT STUDIES

Studies of the influence and variations in the climate upon the human use of the earth contribute in a special way to what may be called geographical climatology.[3] These studies provide information for answering practical questions; in particular they provide information that can be used to clarify the cause and effect and feedback relationships necessary to develop useful climate models for predictive purposes.

Of particular importance are climate effect studies in the areas of energy, agriculture, water resources, and health. For example, problems of energy supply (and costs) have prompted considerable interest in solar radiation as an alternative energy to fossil fuels, along with feasibility studies relating to solar-house construction. The climatologist has also become more involved in engineering applications since construction and design are highly dependent upon the effects of site, orientation, and microclimatic fluctuations.

Most of the research in agroclimatology emphasizes the relationship between climate variability and the productivity of either the 'human-altered' environments (such as the Iowa corn field), or the more 'natural' environments (such as the grasslands of the merino sheep). The economic and political importance of large-scale weather fluctuations must also be recognized. A key research area is determining the impact of weather and climate on food and fibre yields; this has led to a variety of attempts to model the physiological responses of plants and animals and to develop yield equations.

In medical climatology, studies have usually been carried out by the medical specialist rather than the climatologist. Increased knowledge of air pollution and better health statistics, along with improved climate models of possible climate-health relationships, now make it possible for climatologists and the medical profession to study jointly in some detail such things as fluctuations in respiratory diseases, epidemics, and the pattern of hospital admissions, and so learn how climate stresses affect bodily functions.

The importance of climate on human efficiency has also been the subject of many investigations. For example, Bandyopadhyaya[4] in his 1983 book *Climate and World Order* comments on the controversial subject of the impact of climate on 'under development'.

> In India and other tropical countries I have noticed farmers, industrial labourers, and in fact all kinds of manual and office workers working in slow rhythm with long and frequent rest pauses. But in the temperate zone I have noticed the same classes of people working in quick rhythm with great vigour and energy, and with very few rest pauses. ... I had no doubt at all in my mind that the principal explanation lay in the differences in temperature and humidity between the two climatic zones.

C. ECONOMIC AND WEATHER / CLIMATE DATA SOURCES

1. The Data Problem

Attempts to find associations between nationwide economic and weather/climate data present many problems. In particular there is the incompatibility of the two raw data sets since economic information is usually related to areas such as countries or states and weather and climate data to specific sites. It is therefore necessary to transform one of the data sets and the weather and climate data are generally easier to adjust. The economic data banks that do exist are, however, only of limited value in assessing weather and climate impacts, since most economic data is either too embracing in its composition, or it is for large areas or relatively long time periods.

For example, official economic data on the day-to-day sales of say ice cream or antifreeze for a small area is usually not assessed, and it is also generally difficult to find any official economic data available sufficiently soon after an event to be useful in most operational decision-making. Some economic commentators may suggest otherwise, but except for a few notable economic indicators such as commodity and share prices, and electricity consumption, the availability of economic information in any decision-making time framework is minimal, and cannot be compared with the many thousands of weather observations made every few hours at places all over the world and available to meteorological decision-makers within minutes of the events being observed or measured.

2. Sampling the National Scene

Although the traditional base for statistical purposes in many countries is the county or state, in a few countries a limited number of indicators of national economic activity are compiled using sampling techniques. For example, *Business Week* publishes national weekly data on items such as electric power consumption and intercity truck tonnage. In these cases it is important to appreciate that such nationwide data, when used in any weather impact study, needs to be associated with appropriate commodity-weighted weather information for the country as a whole.

That is, national economic data obtained through sampling should not be reduced to a spatial scale that is less than the nation as a whole. For example, data on U.S. intercity truck tonnage are designed to give an index of national activity and should not be considered as estimates of regional activity, nor are the national data a summation of regional data, but rather a carefully designed sample to give a national picture.

3. The Use of Commodity-Weighted Data

Few countries produce and market commodity-weighted weather and climate indices. From both a political and economic point of view this is both surprising and in some ways disturbing, for if climatologists believe that they have a product worth marketing, the assessment and promotion of commodity-weighted weather and climate indices should be actively encouraged. Commodity-weighted weather and climate indices are however available in a few countries, and in New Zealand they are compiled operationally for 120 different weightings (from human population to skiing), and for a wide variety of climatic parameters (rainfall, rainfall moisture deficit, etc.). The use and application of such information is discussed in detail in Chapter VII.

D. WEATHER AND CLIMATE INFORMATION

1. The Value of Weather and Climate Information

A few astute decision-makers often ask the question: 'What is the value (both positive and negative) of weather to a large area (e.g. the United States) over a short period of time (e.g. a week or a month)?' Answers to such questions are difficult, and many climatologists regard any analyses to answer such questions as having little point, since any relationships that might be so derived cannot be used until reasonably accurate weekly and monthly weather forecasts are available. In 1973 when the author first commented on this problem[5] it was assumed that such forecasts would be available within the next decade and that decisions based on such forecasts would or could be made. However, in the light of the now 'recognized' value of real-time weather *information* on items such as soil moisture, heating degree-days, and frost, as distinct from the 'still to be recognized' value and impact of an accurate and reliable medium and long-range forecast, the emphasis which used to be placed on such weather forecasts may not now be as justified as it was first thought.

A distinction should be made however between the rapidly improving 'track record' of 'atmospheric circulation' forecasts, and the poorer 'track record' of the word weather forecasts. Indeed, the 'correct' translation from the forecast 'atmospheric circulation' to the weather resulting from this circulation has yet to be achieved to the safisfaction of many users. It is evident, however, that to many people the value of weather and climate information *has* increased, particularly in regard to the key sectors of food production and energy consumption, and also in explaining - at least in part - economic and political vagaries both nationally, regionally, and internationally.

Until the 1970's the role of the meteorologist in the use and value of weather information, as distinct from weather forecasts, was usually quite conservative, but it is now becoming recognized that the meteorological community (government and non-government) not only must give a strong lead in this area, but also must make sure that decision-makers know how to use such information. The need for this lead cannot be overemphasized, as regretably many top-level decision-makers think of meteorology and climatology only in terms of the weather forecast - and usually only in respect of the weather forecasts that are wrong.

2. Presentation of Weather and Climate Information

The presentation of weather and climate information is a highly specialized and important form of communication. A large number of people are called upon to make weather and climate- sensitive decisions; meteorological and climatological organizations endeavour to satisfy these customers who in general can be divided into 'consumers' or 'households' who use weather and climate services to maintain or increase health, safety, and general welfare, and 'producers' or 'firms' who are concerned in accruing benefits to their 'firms' through the use of weather and climate services in their professional, commercial, or industrial activities.

Many national meteorological services provide weather services to a wide variety of users such as temperature/humidity probabilities, snow warnings, fog forecasts, heavy rainfall forecasts, lightning suppression probabilities, seasonal forecasts, weather-based agricultural production forecasts, frost warnings, rainfall probabilities, wind forecasts, and 'week-away' weather forecasts. Nevertheless, it is fair to say that in general, weather information provided to the public has not greatly improved in either its presentation or impact. Further, since there is an inevitable change through time in the reaction of the public to weather and climate information, it is most important that atmospheric scientists are aware of society's changing desire for, and reaction to, weather and climate information.

3. Weather and Climate Information and Decision-Making

Although it is often suggested that the provision of better information (both weather and non-weather) will lead to better decisions, a great deal depends on the way in which the information is presented to, and used by, the consumer. The presentation of the 'package' to the consumer is critical to the success of the marketing strategy of both government and 'private' meteorological services, and there is strong evidence that this aspect of the 'weather business' will become much more important in the future.

In a recent review assessing user requirements and the associated economic values for both short-range weather forecasts and current weather information the comments of Murphy and Brown[6] are significant:

First, it should be recognized at the outset that requirements will vary from user to user, even for individuals involved in essentially identical activities. ...

Second, the degree of sophistication of users - or, more accurately, the degree of sophistication of their information-processing and decision-making procedures - has important implications for the manner in which user requirements can and should be met. ... Third, studies of user requirements should be sufficiently detailed to ensure that it is possible to differentiate between activities that are *weather sensitive* and activities that are *weather information sensitive*.

Many operations that are affected by weather conditions may not therefore be sensitive to weather information, because of constraints on the options available to the user, or because weather information with the required accuracy or with sufficient lead time cannot be provided. Further, failure to distinguish between activities that are weather sensitive and those that are weather information sensitive, can lead investigators to misjudge the requirements, and therefore overestimate the value of meteorological information.

The number of decision-makers interested in weather and climate information varies considerably. In the United States, the number of 'general public' decision-makers using weather forecasts might vary from a few million to over 100 million depending on the extent of weather activity. But all decision-makers are not equal. For example, the day-to-day economic value of weather-related decisions in the United States range from the 100 million or more who use weather information for convenience only, which may have a value of only a dollar or less, to a few very large decision-making organizations where weather and climate information could have a value in the millions of dollars.

4. Realizing the Value of Weather and Climate Information

In most countries, changing social patterns, trends in population growth, demand for energy, and the impact of marketing have created a need for more rational and effective responses to atmospheric events. The global expenditure by governments on meteorology is estimated to be in excess of $US 12,000 million. It is therefore relevant to ask three questions: what is the current and potential value of weather and climate services, how will advances in the atmospheric sciences improve the usefulness of such services, and what are the potential benefits or losses associated with deliberate or inadvertent weather/climate modification?

Further, in considering 'weather and climate packages', what is the value of weather forecasts published in newspapers, aviation weather forecasts to commercial airlines, or forecasts of extreme weather to hydrological and energy authorities. Moreover, how does the value of this information compare with that which *could* be provided with present technology and finance, and the information which could be provided *if* technology and finance were enhanced?

Answers to most of these questions are not known, and in particular, we do not know how efficient the present 'weather and climate package' is, or whether it is economically desirable to provide more weather and climate-related information. Further, if a better 'weather and climate package' was made available, would users know how to use it, and would they want to use it *and pay for it*?

Managing a weather- and climate-sensitive aspect of an economy or a business must take into account the decisions made by many people with a wide spectrum of responsibility, experience, and training. But the real potential value of the flow of weather and climate information in this context can be realized only when it affects decisions taken at the top level. In this regard the viewpoint first put forward in 1972 by McQuigg[7] is to the point.

> ... In some important instances, the greatest contribution a meteorologist can make is to find a way to establish real communication between a certain group of decision-makers and the meteorological system. This is not to say that all a National Weather Service has to do is to provide information to management (or to the general public), and then wait for the economic benefits to become obvious to all concerned. But it surely is true that no economic benefits will accrue to the meteorological information system from a weather-sensitive segment of the economy in which the persons making management decisions are (1) not aware of the availability of meteorological information; (2) aware of the potential value of meteorological information, but do not have any channels through which such information may be received.

E. BENEFITS AND COSTS OF WEATHER / CLIMATE INFORMATION

1. Actual and Potential Aspects

An important aspect of the weather and climate sensitivity of various sectors of an economy is the actual, realized, or potential value of information. The relationship is not simple, and in many cases it is easier to assess the value of weather and climate information to a highly sensitive but very small economic sector, rather than a slightly sensitive but relatively large sector of an economy. For example, although the raisin industry[8] in the United States is relatively small economically, it is highly sensitive to some kinds of weather and climate information which therefore have a potentially high value. In contrast, the many aspects of manufacturing,[9] although relatively high in overall economic importance, usually have a low sensitivity to weather and climate information and therefore have a relatively low value.

Several national attempts have been made to place a dollar value on weather and climate information including weather forecasts. Some of the first such analyses[10] showed a benefit/cost ratio varying from about 5 to 1 to about 20 to 1. It is extremely difficult to calculate in detail the various benefits (particularly those associated with 'freely' available public weather forecasts), but it is evident that the costs of providing weather and climate information in most countries is relatively small compared with the potential benefits. Very few in-depth studies of the actual benefits to various sectors of an economy have been made, and the need to refer to studies made more than 20 years ago point clearly to the dearth of 'hard' information on this subject.

2. Agricultural Benefits

The annual benefits of weather forecasts to agriculture in the United Kingdom were provisionally estimated in the mid-1960's to be at least 0.5% of the gross farm income.[11] Translating this percentage to an agricultural country like New Zealand would give a probable benefit (in 1988 dollars) of about $US1 /farm holding per day. Such a value will vary from area to area and country to country depending on what weather and climate forecasts are available, how they are used, and how much of their potential value is being realized. The question of assessing the value of climate forecasts (as distinct from weather forecasts) is equally difficult, but it is estimated that climate forecasts one, two, or three months ahead would have a *potential* value considerably in excess of those suggested for weather forecasts. The value of real-time and historical weather and climate information (as distinct from forecasts) also needs to be assessed, and with the advent of electronic information systems, the potential is considerable.

3. Building and Construction Benefits

Weather losses in the building and construction industry in the mid-1960's were estimated to be 2% of the value of production in both the United States[12] and the United Kingdom.[13] It was also assessed that about 10% of these losses could have been avoided through the appropriate use of the weather forecasts available at that time.

The specific use of weather forecasts at construction sites in Sweden has also been examined, Wahlibin[14] finding that while the current day's activities are very much predetermined, plans for the coming day such as decisions to cast concrete rank high and are clearly dependent on the forecast weather. The economic value of weather forecasts to construction in Sweden was also assessed, and it was found that in relation to total turnover, the estimated value of forecasts was 0.1 to 0.2% of the total costs. This was about seven times the price paid for the forecasts.

4. National Benefits

The total benefits of weather and climate information to any nation include items in addition to those just discussed. In particular, the energy sectors in most countries are not only large but usually very weather sensitive in terms of requirements for heating, cooling, transportation, and agriculture. There is also the value of public weather forecasts and the more specialized weather and climate information used for such things as ship routing, advertising, and highway maintenance.

In New Zealand, the total value of weather and climate information in 1968/69 was provisionally estimated to be NZ$32 million, compared with the then costs of providing these services of NZ$1.8 million.[15] A more recent analysis by the author indicated the benefits to agriculture, building and construction, and manufacturing to be NZ$80 million which, compared with the then annual expenditure (in 1985) of NZ$16 million, gave a benefit/cost ratio of 5:1. But if all sectors of the New Zealand economy were considered the benefits would be considerably higher; therefore it is reasonable to suggest that in most countries an analysis of the benefits of weather and climate information to all weather and climate sensitive sectors would provide a benefit/cost ratio of at least 5:1, and more likely a ratio exceeding 10:1.

A common theme to most analyses on the benefits of weather and climate information is the difficulty of obtaining information as to the real value of weather information,[16] and the comments by Mason in 1966[17] in this regard are still pertinent:

> One cannot apply commercial criteria based on the price the customer is willing to pay because the basic service is free. The economic value of a special service for a particular customer can usually best be judged by the customer himself, but he may be reluctant to divulge this for fear that the charges will be increased. In some cases it is sensible to ask to what extent would a particular industry suffer in the absence of a weather service, but others, such as aviation, could hardly have become established in these circumstances.

F. RESPONSES TO WEATHER / CLIMATE VARIATIONS

Although human activities are 'sensitive' to weather and climate variations on a wide range of scales, it is at times extremely difficult to assess these relationships. Difficulties arise first because little research has been done on assessing climate sensitivity, and second, what research has been done is often regarded by the critics as being either superficial or obvious. But while the concept of sensitivity may be obvious, a measure, or more importantly a scale to measure these sensitivities, is not obvious. For example, is the raisin industry of Australia more or less sensitive (to the weather and climate) than say coffee production in Brazil, or the area of wheat sown in the Soviet Union, or the quality of wool production in Uruguay, or the tourist industry in Jamaica?

Developing nations are usually considered to be more sensitive to weather and climate than the developed nations. The true situation however is probably the reverse, for while developing nations may indeed be more vulnerable, the sensitivity of developed nations is usually more important in an 'absolute' sense. For example, wheat production in Canada may not only be more sensitive (in the economic, social, and political sense) to a weather and climate variation than say rice production in Thailand, it may also be more important in absolute terms. However, few studies have shown the true 'value' of weather and climate sensitivity, and with the present knowledge only an educated guess or a 'feel for the situation' is possible.

G. COMMUNICATING WEATHER AND CLIMATE INFORMATION

1. Communications

In several countries there are specialist information companies who can provide users with any kind of weather and climate information they require - at a price. Most of this information is highly time dependent but whether the question relates to, say, the optimal routing of a ship from Sydney to San Francisco, or weather and soybean production in Brazil, information about the past, present, and future weather and climate conditions *is* available to assist the decision-maker.

The presentation of such information is a highly specialized and important form of communication, and the weather and climate package can be viewed in three phases: the preparation of the package by the meteorologist or climatologist; the presentation of the package through radio, television, newspapers, journals, or videotex systems; and the use or application of the package by the consumer. To market the weather and climate package to the consumer effectively, meteorology and climatology needs to broaden its vision by actively encouraging research in the applicable social, economic, and marketing aspects of the subject. This could be done by encouraging meteorologists to look anew at some of these problems. Employment of social scientists, economists, and marketing experts in meteorological services should be given particular attention, and the actual sponsoring of research and weather information services should be developed.

2. The Popular and the Business Press

The *Press* (Christchurch, New Zealand), in an Editorial on 10 October 1981, stated that '...the time is long past when weather forecasting was something of a national joke and when the predictions of meteorological services were regarded - no doubt unfairly - more as an expression of opinion than as scientifically based forecasts.' However the media are not always so constructive or astute - indeed the *Evening Post* (Wellington, New Zealand) in an editorial *Weather Watch* on 22 August

1981, with reference to the new computer at the British Meteorological Office, stated:

> Hope springs eternal! The new computer ... will at least be able to tell you that it's useless to waste 6 days at Lords ... for the one hour cricket play possible each day ... come to think of it, this could have an electrifying economic effect. Company directors ... instead of lounging around the Long Room ... may actually be doing some work. Perhaps there really is method in the British Met. Office's madness!

The 'weather game' the media play is quite fascinating and sometimes entertaining, but there are also some influential newspapers and business publications that clearly have a much greater awareness of what it is all about. In recent years, editors of several influential economic, news, and planning publications have become more aware of the importance of weather information in national and international business activities and its growing importance in national planning. For example, *Newsweek* (2 January 1984) stated:

> As ever, human error and aggression, flukes of nature and plain bad luck left ordinary people suffering. ... Even the weather seemed especially bad around the globe, apparently because of the shifting of the Pacific current called El Nino. Drought and famine afflicted large areas of Africa, Australia and Brazil, and floods washed out homes and farms in Europe. And in the United States, a heat wave withered some of the world's best grain fields, sending clouds of priceless topsoil blowing in the wind.

The human aspects of climatic uncertainties are usually well reported in the media, but a key aspect of the daily newspaper in communicating weather and climate information which should not be overlooked is the daily weather column,[18] which provide 'all that editors believe you need to know' about the local, regional, and global weather scene. There appears to be a wide difference of opinion as to what should be published, but mention should be made of the enlightened viewpoint of the management of *U.S. Today* which publishes a full colour page of weather information in its European and Asian editions, as well as a full colour page in its United States edition of the domestic weather scene, plus informative comment, description, and/or diagram of a meteorological event of social, economic, or political importance.

3. Television Weathercasting

The 'Weather' is a prime television programme, and in many countries it is deliberately put at the *end* of a news programme to hold an audience. In some countries the role of the weathercaster is fairly basic and conservative especially when compared with the ' Weather' television channel in the United States which operates 24 hours a day through a cable television network. However, whatever type of weathercast is available, viewers are always keenly interested in how and what is presented, and an essay in *Time* (17 March 1980) on 'The Wonderful Art of Weathercasting' aptly describes weathercasting as '...an odd and specialized calling: not exactly journalism, not exactly meteorology, not exactly soothsaying, not exactly show business, but parts of all four.'

But it is also more than this and viewers have very definite views of what they like and dislike. The 'art' of weathercasting has had a long period of development, and in the United States a television company will spend many hundreds of thousands of dollars to attract the 'right' weathercasters, and provide the 'right' image' for the viewer. Nevertheless, there can be too much of a good thing, and the following extract from the *Evening Post* (Wellington, New Zealand) on 24 August 1985, written by Francie Brentzel, highlights the fact that 'getting the forecast right' is presumably what it is really all about.

So I said to him, 'New Zealanders just don't know how to "mediatize" the weather like in the States'. 'What do you mean?' he said. 'Well, last night this very distinguished-looking television news reporter said, "We are forecasting fine weather in Wellington tomorrow" '. 'What's wrong with that?' he said.

'What's wrong! That was it. That's all he said. I want to know the average temperature for this date and the all-time high and low temperatures for this date since biblical times. I want to know high tide and low tide, the exact moment of sunrise and sunset, and the phase of the Moon. I want to know the barometric pressure, is it rising or falling, and if there are any small craft warnings.'

'Why do you want to know that?' he said. 'We don't have a boat.' 'Don't interrupt. I want to see the weatherman point to cities on an electronic map; then I want the satellite still-photos of New Zealand speeded up to simulate weather systems' movements across the country. I also want to know the humidity, the pollution index ...' 'Look,' he said, 'the man said it was going to be fine today in Wellington and it certainly is fine, isn't it?' 'It's just that I'm used to useless facts.'

4. Videotex Systems

Videotex systems are designed to provide information to specific groups such as wheat growers, grain storage companies, kiwifruit growers, urban planners, international soybean buyers, hydro-electric authorities, railway operators, television companies, and wool marketing companies. The type of information that can be provided to these specialists include the whole range of weather and climate products, including - at least in theory - every weather and climate forecast, and every piece of weather and climate data available in every national meteorological service. That such a range of information will be available to any customer, at the touch of a button, creates a formidable marketing and educational challenge to meteorologists and climatologists, which will be realized if a bold and imaginative approach is used.

WEATHER MONITORING AND FORECASTING

A. MONITORING AND UNDERSTANDING THE WEATHER/ CLIMATE RESOURCE

1. Observing the Weather

A vast and complex weather reporting network is an essential component of many national meteorological services. A good example of how vast a network is required is provided by New Zealand, geographically isolated and small in area, but which gathers weather data from stations as far north as the Equator and as far south as the South Pole, and in an area extending to the west of Australia and east to Easter Island, as well as its own land-based stations.

Some island stations outside New Zealand (e.g. Raoul and Campbell) are fully manned by New Zealand Meteorological Service personnel, while others through-out the South-West Pacific are administered from New Zealand. Reports from this network of stations are part of the World Weather Watch Programme, organized by the WMO. Other national meteorological services have observing networks of varying size, and all rely to a considerable extent on weather observations from the global weather community of nations.

The kind of weather observations made vary from country to country in terms of their number, spacing, elements observed, and their frequency of observation, but the situation in New Zealand where there are basically three kinds - rainfall, climatological, and synoptic weather stations - is fairly typical of most countries. In New Zealand 1,600 rainfall stations send in by mail monthly reports, and 350 climatological stations provide more detailed information including sunshine duration, wind, cloud cover, pressures, temperatures, etc. Most of this information is not used in a real-time sense for forecasting, but is used rather for compiling background and day-to-day reference-type climate information to give an overall picture of present, recent past, and past weather and climate patterns.

As in all countries, however, the synoptic weather stations along with observa-tions from ships, floating buoys, aircraft, and other 'remote sensors' are the key to day-to-day forecasting and provide real-time weather information. Reports every three to six hours are available in New Zealand from up to 150 synoptic stations (some are automatic), and hourly reports from up to 30 of these stations. In addition, most New Zealand synoptic stations provide with their 9 am report information for the previous 24 hours on the rainfall, sunshine, radiation, soil temperature, wind run, and maximum and minimum temperatures, and this forms the basis for a real-time weather information service.

2. Climate Data Sources

While traditional climate data archives provide the basis for a large amount of climate research and climate information activities, there is a need to extend or in some cases relocate existing data-collection networks so that the available observations provide the most representative user-orientated data. New sources of data also need to be considered, and observations from satellites which provide information on items such as the earth's radiation budget, cloud distribution, and atmospheric composition need to be included in any catalogue of data sources. It is also important for climate indices to be developed that can be used in climate models, and as previously noted there is a particular need for commodity-weighted climatic indices to be available for political or 'administrative' areas, and for relatively short-time periods, for use in real-time decision-making.

In the past climatologists examined available long-term records of instrumental observations of temperature and precipitation and evaluated them for consistency and accuracy. Spatial coverage of such data is usually very poor however, and only a very careful use of historical documents provides a means for obtaining data beyond the most populated and longest-settled regions. An example of such work is the use of the 'content analysis' technique to journals of the Hudson's Bay Company to determine the dates of annual freeze and breakup of estuaries along Hudson's Bay between 1714 and 1871.[1]

Many national meteorological services are making a significant effort to improve the completeness of operational weather and climate data. Efforts to obtain such data from ocean and remote land areas, to improve the usefulness of data from satellites, as well as to improve the management of the data collected, are necessary to provide world-wide, consistent, compatible climate data in a timely manner. Internationally the Global Atmospheric Research Programme (GARP) of WMO represented the first real international effort to collect a world-wide data set defining the state of the global atmosphere. The data collected have been used to test and improve our understanding of the global climate, and by inference our ability to forecast the weather/climate resource in the medium range (a few days to a few weeks) and the longer range (a few weeks to a few months).

The collection, processing, verification, and organizing of climate data from so many sources presents a major challenge. But it is also important to understand the economically desirable and attainable levels of observational accuracy and spacing, including the use of both manned and automatic observation sites, the appropriate parameters which need to be monitored, and the special attention that needs to be given to satellite systems for monitoring sea surface temperature, soil moisture, and aspects of the radiation budget.

3. Climate Change and Climate Variations

'Climate change' is paramount in any consideration of weather and climate as a resource, but several aspects of the term need to be understood before any worthwhile analysis of the problem is considered. These include: what is a climate change, how is the climate changing, what are the possible causal mechanisms, can these mechanisms be modified by the human environment, are the effects of a climate change predictable, and what are the economic, social, political, and strategic implications of a climate change?

Although several terms are used when describing climate change such as 'variations', 'trend', 'oscillation', 'periodicity' there is little uniformity in their use and acceptance. For this reason, it is perhaps desirable to 'define' climate or climate state as 'the totality of weather conditions existing in a given area over a specific period of time'. Climate change could then be said to become important only when a relatively long time period is considered. But such a viewpoint is too restrictive. Indeed the importance of monitoring, analysing, understanding, and (where possible) forecasting climate variations over *all time periods* from days to decades is of far greater importance, whatever the 'true' meaning of climate change may be. It is also necessary to understand the increasing impact that a climate change (however defined) is likely to have on economic, social, political, and strategic activites.

4. Monitoring Climate Variability

The need for more comprehensive monitoring and analysis of the world's climate to detect and predict changes, and to understand the consequences of such changes, is central to many aspects of national planning. Sewell and MacDonald-McGee in discussing the climate scene in Canada[2] stated:

> At present...understanding of the manner in which climate affects human activities...and the ways in which the latter may cause changes in climate are not well understood. Despite a recognition of this deficiency by various (people) progress in mounting research has been very slow. Problems associated with the management of a hitherto unrecognised resource, climate, seem certain to become a major focus of policy formulation before the end of the century.

Regardless of the political, economic, strategic, and social impacts of any long-term trend, it is evident that year-to-year variations of considerable significance will continue to occur, and it is important to re-emphasize that trends in the climate over time - *to which it is possible to adjust* - are usually relatively small by comparison with short-term variations. Also of importance is that most political regimes have a 'thinking lead-time' of only a few years; accordingly, the economic and political significance of a longer-term trend in the climate is often difficult to infiltrate into many political decision-making processes. But in some cases they do infiltrate the influential press, as is evident from this comment in the *Financial Times* of 3 November 1983.

> An air of certainty has crept into recent projections of how the global climate is going to change in the future. The majority view is that the general trend is firmly in the direction of a warming of the climate. Only a decade ago .... the consensus was equally determined that we were heading for a new ice age. ... So why the sudden change, and can we place any more faith in the latest set of forecasts?

While these comments may be typical of how the press see the work of meteorologists and climatologists, the implications are clear that it is time for climatologists to put 'their house' in order. As part of the World Climate Programme, WMO is assisting in this process, and Unninayar[3] highlighted some of these concerns when he noted that measurements taken over a period of time, on a global basis, enable the construction of the history of climate, studies on climate trends and climate variability, and the interactive processes which are involved in causing such changes. He further commented:

> As a priority requirement it is considered necessary that a mechanism be formulated, relying on existing research/analysis/operational centres, to compile and disseminate summarized information on the present status of the climate system, changes from past years, indication of trends, and significant events of regional and global consequence.

The politician, planner, and decision-maker must also become more weather and climate orientated if optimum use is to be made of the weather and climate resources of the 1990's and the 21st century. An important first step in this regard is to understand more fully the *specific* benefits of weather and climate information to economic, social, and political activities, and at the international level this will be examined in detail at a WMO Technical Conference on 'The Economic and Social Benefits of Meteorological and Hydrological Services' to be held in Geneva in March 1990.

B. WEATHER FORECASTS

1. Balancing Scientific Enquiry with the Value of the Product

Much work has been accomplished on modelling the physical atmosphere; however much less research has been done on the economic and social aspects of

such modelling. Indeed, surprisingly little thought is usually given to the end of
the meteorological chain, a chain which comprises many parts from the basic
observations through to the *user* of the forecast. The key question at the end of this
chain is the value of the product; that is, what is 50 mm of rain 'worth' to an
individual, area, or nation, and what kind of dollar value can be placed on the
overall benefits or costs of a correctly forecast dry period of two weeks to say the
farmers of Indonesia or Kenya, or the value of a correctly forecast cold snowy day
to the people of New York, Moscow, or London?

Suppose for example that a major snow storm is forecast to move over London
tomorrow. Do Londoners simply accept this, and adjust accordingly, or do they
convince their politicians and scientists to say divert it towards Paris. In either case,
if (among other things!) the impacts of a snow storm on London (and presumably
Paris) are known, then we are in a better position to decide which action to take.
But do we know?

Answers to such questions are important because of three factors: (1) if the
current use of weather forecasts is not known, then any improvement in the
weather forecasts - from the user's viewpoint - may be not only a waste of time,
but also a waste of public funds devoted to them; (2) more effective *use* of weather
forecasts may well be more useful than an improvement in the accuracy of the
weather forecast; (3) if we know what tomorrow's weather *and its effects* will be,
then we are in a better position to alter our activities, or call upon the services of
the weather modification expert.

2. The Weather Forecast: Its Use and Value

For many years there has been considerable research into long-range weather
forecasting, but with limited success, indicating that there is no quick or easy
solution to the problem.[4] The apparent need by the public and specialist users for
longer-range forecasts is however still strong in spite of the fact that many
decision-makers in weather-sensitive industries would not be able to use them
because they don't specifically know how the present and past weather affects their
operations. For example Glantz,[5] in assessing in 1976 the value of a long-range
weather forecast for the West African Sahel, indicated that although a reliable
long-range weather forecast is not yet possible, it may not even be desirable for
many parts of the world until that time when some essential adjustments to existing
social, political, and economic practices have been undertaken.

A key question is 'where do other nations fit into this?' Have the essential
adjustments to existing social, political, and economic practices been undertaken?
In most cases the answer is no; indeed many businesses and governments seem to
be quite content to ignore both the effect of the present and past weather on their
activities, as well as the impact of the future weather with a consequential loss of
profits to their 'shareholders'.

3. Weather Forecasting: Analysis into Words

What is involved in making a weather forecast? Let us consider - by way of an example - the first national forecast of the day, issued by the New Zealand Meteorological Service at 5.30 a.m. This forecast is based on surface and atmospheric observations from a network of reporting stations, as well as on satellite pictures received during the night. The forecaster begins by analysing the most current weather map, compares it with previous ones, and make a prognosis (a projected weather map usually for 24, 36, or 48 hours ahead). In addition several computer-based predictions of the state of the atmosphere at specific times in the future are used, these 'numerical weather predictions' being particularly important for regions in which there are few land or ship observations.

On the basis of the various prognostic maps, one final version is produced, and the forecaster then produces the *word* weather forecast (rather than more lines on a map!). But the 'human touch' is still very important, for although a computer model may quite accurately predict the general weather situation, such as the location of anticyclones (highs), fronts, or depressions (lows), the skill of the human forecaster is needed to interpret this in terms of the weather to be expected by the user. Local variations in weather patterns are a very real problem, and experience, research, good local knowledge, and the gift of communicating with the user are extremely important.

Computers also allow high-speed processing of all incoming data, including satellite signals, and they permit the high-speed communication of forecasts and real-time weather information. Efficient communications of both inward and outward information is in fact very important. For example, in terms of outward information, the New Zealand Meteorological Service each year issues 66,000 radio and 20,000 newspaper forecasts (for its 3 million people and 65 million sheep!), 78,000 marine forecasts, 56,000 international aviation forecasts, 72,000 local aviation forecasts, and 2,000 television forecasts. In addition, over 200,000 individual telephone calls are received from customers requiring specific weather and climate information, and over 8 million telephone calls a year are received for weather forecasts on the automatic 'metphone' system.

4. Weather Forecasts and the User

Although improvement in the accuracy of weather forecasts is the prime object of meteorologists, as has been mentioned before too often the translation from the weather forecaster to the consumer is not given the same care and attention as the translation from the weather maps to the forecaster. Some attempts have been made to redress this situation, and the rainfall probability forecasts used extensively in the United States are a good example of translating the weather into terms which the user is familiar with.

But there are still many aspects of weather forecasting of which we are generally ignorant. For example: How many people read the weather forecast in their local newspaper, listen to a radio weather forecast, or look at weather on television, and *remember what they read, heard, or saw?* How often are the real 'thoughts' and doubts of the weather forecaster transferred to the user of the forecast? What effect does a weather forecast have on the user, how many users reschedule their activities because of the forecast, and does it make any difference whether the forecast is accurate?

During the past three decades there has been a significant improvement in the 'accuracy' and usefulness of weather forecasts, and a greater awareness of the demand for weather forecasts by a wide variety of users. But it is fair to say that *to the public*, weather forecasts have not improved in presentation or impact, and a change over time in the reaction of consumers to weather and weather forecasts is inevitable. It is therefore important that weather forecasters are aware of the public's understanding of and need for forecasts.

A survey to assess answers to some of these questions was made in New Zealand in 1984 when a 'face-to-face' interview of 1,000 farmers was conducted by AGB: McNair Surveys Ltd, a professional client and consumer organization specializing in social and marketing research. Two principal questions were asked; one related to the preference for one or other of the two weather presentations on New Zealand television, the second asked about the source (e.g radio, television) of weather forecasts and their relative importance.

The results showed that overall there was equal preference for the weathercasts presented by forecasters from the Meterological Service, and those presented by contract staff (non-forecasters) of Television New Zealand, but with marked regional variations. The differences (as much as 66% for and 69% against a particular presentation) were partly related to the type of farming carried out, as well as the time of presentation, and whether it was part of the 'main' television news programme.

Answers to the second question showed that 'specialized' weather forecasts, such as the 4-5 day forecasts broadcast over the 'quality' nationwide (non-commercial) radio network, attracted only 20% of the possible audience (similar to the overall radio listening preferences of New Zealanders), 72% viewed either of the two television weather presentations, 46% listened to weather forecasts on their local or community radio stations which in New Zealand are supported by advertising, and 22% obtained weather information from their morning newspaper.

A number of interesting factors emerged from the survey, the main one being that - at least in New Zealand - even specialized users, such as farmers, are fairly traditional in their source of weather information and that despite the best intentions of the New Zealand Meteorological Service, most do not listen, watch, or read the specialized and/or professional presentation of the weather, preferring instead the more 'popular' presentations on their local (commercial) radio station. Nevertheless, the survey showed that there are a significant number of people who do want the more specialized and professional presentation of the weather, and it is important that the 'market' significance of this group of 'more informed customers' continues to be recognized.

All user surveys of the requirements for weather forecasts indicate that time may alter the consumer's understanding of weather forecasts and terminology, and also the user's requirements for weather forecasts. More study along these lines may in fact be just as worthwhile in the long run as more accurate forecasting, however desirable the latter maybe.

5. The Value of Weather Forecasting: Future Aspects

If tomorrow's weather, *and* what we do because of it, is to be more a matter of choice than chance, more research on the process of decision-making in the management, use, and possible modification of the weather is needed. Studies are required to identify what weather-related decisions are made, who makes them, and what factors appear to influence their outcome. For example, what decisions are made about adjustments to a forecast snowstorm or a forecast severe windstorm by individuals, private industry, and various levels of government?

Research is required on how an increase in the amount of information, or an improvement in the accuracy of weather forecasts, can lead to changes in production schedules or alterations in human activities. Further, to what extent is the meteorologists' view of the value of increased weather information borne out by the manner in which people and companies *actually* use the information provided? Knowledge of these matters is clearly of considerable value to those involved in designing weather forecast 'programmes' as well as to those developing policies to encourage more efficient adjustment to weather variations, and the more efficient use of weather forecasts.

IMPACTS AND SENSITIVITIES

A. CLIMATE SENSITIVITY

1. An Overview

Climate variations - including 'changes' in the climate - are (as already noted in Chapter 5) paramount when considering weather and climate sensitivity, but many aspects of these terms need to be considered. They include answers to the questions: what is climate, what is a climate change, what is a climate variation; is there real evidence for a climate change or climate variation in terms of specific changes in temperature, rainfall, soil moisture, etc. If so, in what way is the climate changing; what are the causal mechanisms and can these mechanisms be modified by human activities; are the effects of a climate change or variation predictable, and at what degree of accuracy; and what are the implications (economic, social, strategic, and political) of climate variations and climate change?

Many weather and climate variations clearly have significant economic consequences, but since they also affect human activities in many other pervasive ways, it is to be expected that people should try to find ways of adjusting to and even altering the weather and climate. It is also important to recognize that modification of tomorrow's (or next month's) weather and climate should be dependent on the weather and climate that is forecast to occur, particularly if decisions to modify are based on what is predicted. Moreover, to the user, accurate weather and climate forecasting and successful weather and climate modification may have similar impacts. For example, the retail store manager or the cattle farmer is not concerned with *why it rains*, but only that it does or does not rain.

It is equally important to appreciate that more accurate longer-range forecasts, and the ability to modify tomorrow's weather, can have disadvantages. This is because if *everyone* knows what next week's or next season's weather is going to be, and there is no control over its possible modification, then there could be conflicts resulting from the different needs of people. There will also be no economic advantage to those who 'pay' for information about what is going to happen in the future.

2. Characterizing Climate Sensitivity

There is no single method for characterizing a society's climate sensitivities, but several steps might be considered, including an analysis of the uses and users of weather and climate information. This would provide an idea of which groups or activities value climate knowledge. A review of the appropriate literature related to climate impacts could also be made, which would lead to ranking activities as to their sensitivity to climate influences.

Whatever methods are used, the climate impact assessor needs to complete an assessment of relative or absolute sensitivities, and then express these values in a usable form. One approach is the calculation of quantitative indices that combine the climate state and the economic activity into a useful 'single number'. This 'single number' can be thought of as an 'econoclimatic indicator'. It should be noted, however, that the typical response of many people to interpreting such regional and national economic indicators is usually very conservative - or more kindly, just not understood. Such econoclimatic indices can now be computed and made available regionally, nationally, and internationally, and if used with economic and political understanding, can give very useful decision-type information.

3. Identifying and Assessing the Sensitivities

The whole range of weather and climate services depend upon an in-depth knowledge of weather and climate sensitivity, but what activities are sensitive, and what part do weather and climate variations play in these activities? Consider for example the following: (1) how much variation is needed before it becomes significant in producing effects or in affecting decisions; (2) are such significant weather and climate variations the same from one area to another, one day to the next, one season to the next, and from one type of activity to another; and (3) what specific effect does a significant weather and climate variation have on particular activities such as the restaurant business, tourism, retail sales, unemployment, ice cream sales, airline operations, or gasoline sales?

If we had these answers we would be in a better position to evaluate weather forecasting and climate programmes, those involved in weather and climate sensitive activities could operate more efficiently, the economics of weather and climate modification activities could be more precisely determined, and adjustments to tomorrow's weather and the climate of the next decade would be more a matter of choice than chance.

In assessing an economy's climate sensitivities, basic steps include a review of the research related to climate impacts, and an analysis of which groups value climate knowledge and are exhibiting sensitivity. However, in many nations a more immediate and useful guide to weather and climate sensitivity is obtained from a critical analysis of the relevant agricultural, economic, and business journals, and by a careful appraisal of the nation's newspapers. Indeed in many countries the media often provide the only indication of the importance, in a real-time sense, of the weather and climate. In particular, a media analysis is particularly important in data-poor nations where what little data do exist are available for analysis only months or years after the event, and in those data-rich nations where decisions are made on a day-to-day or hour-to-hour basis and where the prices of commodities are paramount. In both cases the financial, economic, and agricultural sections of the daily press (or specialized newspapers such as the *Wall Street Journal* or the *Financial Times*) publish valuable information provided the reports on the 'economic climate' of sectors, regions, and nations are read with a critical eye.

One specific pioneering study assessing sensitivity was made in 1980 by the Center for Environmental Assessment Services[1] of the U.S. Department of Commerce, in which the sensitivity of elements of the gross national product (GNP) to widespread anomalous weather events was examined. A summary is given in Table VI.1, which is compiled from an analysis of the weather and climate sensitivity of various economic and social sectors in the United States as reported in the *New York Times* during a 10-year period. The survey showed, among other things, that a major increase in personal expenditures on electricity and food at home can result from an unusually hot summer and an unusually cold winter, and that a major increase in personal expenditures on furniture and appliances, and food away from home can result from unusually mild weather.

Specific weather and climate events were also studied by the Center. In particular, during the 1980 summer heat wave and drought in the United States a series of reports updated in near real-time the mounting economic losses that were finally estimated at more than $US20,000 million. Six months *after* the last special report[2] on these events was issued, official statistics were released confirming the earlier weather-based estimates. To compare the impact of weather and climate of this event with previous similar events, a report was also prepared on the 1976/77 winter which indicated that the economic losses were almost twice those of the 1980 summer heat wave and drought. Such weather/media-based studies of a national economy are a very useful method of placing specific weather and climate events in perspective.

4. Review of Weather and Climate Sensitivity Studies

There are two overlapping methods of surveying activities or areas that are highly sensitive to weather and climate. The first are studies of users, and the value of the information that 'national' meteorological services generate; the second involves studies of scientific symposia, or governmental and intergovernmental conferences.

Impact studies relating to 'national' meteorological services include the mid-1960 assessments for the United Kingdom[3] and the Soviet Union,[4] which emphasized the importance of meteorological conditions and information for agriculture, and pointed out their role in a variety of other activities, such as aviation. Later, in 1979, a symposium in New Zealand[5] discussed the value of national meteorological services, the topics including the relation of weather and climate to forest fires,

Table VI.1 Sensitivity* of Gross National Product elements to widespread anomalous weather - United States

GNP elements	hot summer	cold winter	dry summer	storm/rain	snow	mild
1. Personal consumption expenditures						
(a) Gasoline and oil	−	−	−	−	−	+ +
(b) Electricity	+ +	+	?	+ +	?	−
Natural gas, fuel oil, coal	?	+ +	?	+ +	+	−
(c) Furniture and appliances	−	−	−	?	−	+ +
(d) Food at home	+ +	+	+ +	+	+ +	− −
Food away	− −	− −	?	?	− −	+ +
(e) Apparel	−	+	?	?	−	+
(f) New and used cars	−	−	−	−	−	+ +
(g) Housing	−	− −	?	?	−	+ +
(h) Transportation	−	−	?	−	−	+ +
(i) Other	?	?	?	?	?	?
2. Non-residential fixed investment	?	?	?	?	?	?
3. Residential	−	−	−	?	−	+ +
4. Change in business inventories	+	+	+	+	+	− −
5. Net imports	+	+ +	+	+	+	− −
6. Government purchases						
(a) Federal	+	+	+	+	+	−
(b) State and local	+	+	+	+	+	−

Source: After Maunder and Ausubel (1985), adapted from Center for Environmental Assessment Services (1980).
Note: * Weather-related changes in consumption: + = increase; + + = major increase; − = decrease; − − = major decrease.

forest management, wool production, ship operations, electricity supply, the siting of power stations, boating, and sport.

A second major source of information on weather and climate impacts is research literature derived mainly from conferences. The agenda and contents of the reports provide various judgements about what is sensitive to climate. A comparison of these individual agendas[6] with the author's major study in 1970[7] suggests both the convergence of opinion on climate sensitivity as well as the fickle judgements.

An alternative to the identification of the sensitivity of large commodity sectors or regions is to focus on specific topics. A Massachusetts Institute of Technology conference in 1980 on *Climate and Risk*[8] illustrated some of these sensitivities: extreme wind speeds and structural failure risks, weather hazard probabilities and the design of nuclear facilities, impacts, and use of climate information in the hail insurance industry, forecasting for offshore drilling and production operations in the petroleum industry, seasonal climate forecasts and energy management, evaluating farming system feasibility, crop growth models and climate data, and snow management and its economic potential in the Great Plains.

Sensitivity judgements based on studies of the value of national meteorological services or conference agendas have, however, a significant deficiency in that their dominant concern relates to industrialized societies. For example, only a few papers[9] from the 1979 World Climate Conference relate to developing country situations. Thus, although there is evidence that primary activities (agriculture, pastoralism, water resources) *may* be even more sensitive to climate in developing countries, specific surveys of what the real sensitivity is in many such countries remain to be done.

5. Analysis of Information Uses and Value

Several attempts have been made to place a dollar value on weather and climate *information*. Such analyses require looking at a national economy through the 'joint eyes' of a meteorologist and an economist. This means that a critical analysis of national economic data needs to be done in order to rank those parts of a nation in terms of both their sensitivity to weather and climate *and* their contribution to national productivity.

Looking at sensitivities from the viewpoint of the value which both the provider of information and the user of the information places on a particular product, three factors need to be considered: (1) the weather and climate sensitivity of various sectors of an economy and their relative importance; (2) the separation of weather and climate information into weather and climate forecasts, real-time weather and climate information, and past weather and climate information; and (3) the difference between the value of presently and potentially available weather and climate information.

A good example of the thinking required in this kind of exercise is given in the pioneer study of the United States building and construction industry by Russo[10] in the mid-1960's. He estimated that the total weather losses ranged from 1.1% to 11.3% of the value of construction. In terms of the total value of the *weather-sensitive* sectors this represented a range from 2.5% to 25.2%. Russo noted that 'this wide range ... results from a highly speculative estimate of the decreased construction volume due to seasonal weather effects.' This comment highlights the real difficulties in giving anything but approximate values of the weather-related losses of any weather-sensitive sector, and they relate directly to the problems associated with identifying and ranking weather and climate sensitivity.

6. Quantitative Analyses

An initial quantitative approach to identifying climate sensitivities might be to look for seasonal variations in production and consumption data. For example, many retailers are aware that there will be a significant increase in buying by the public in early spring and early winter, whereas a contractor orders materials and hires additional workers for the increased construction activity that inevitably comes in many countries during the summer months.

A statistical correlation analysis may also suggest sensitivities when climate and economic data are compared. For example, a study of weather and the sales in autumn of women's winter coats in New York departmental stores was made in the early 1960's. The analysis[11] showed that about 15% of the September-December sales of coats occur in September when the September average temperature is about 23°C (73°F), but that 20% of the same sales occur when the September average temperature is about 18°C (64°F) or 5°C (9°F) colder.

A good idea of climate sensitivity can also be obtained from an analysis of appropriate climate and economic data of a country. Such an analysis for the New Zealand sheep and dairy industry is discussed in detail in Chapter X. Similarly, the impact of a severe winter and a hot, dry summer on British industry was assessed in an analysis by Palutikof[12] who showed that severe winters increase activities in the utility industry by 17.5% compared with the same month for a 'normal' season, but reduce clothing and footwear production by 9.6%, while a particular summer drought reduced utility activities by 6.4%, but produced increases of 3.0% in clothing and footwear production.

While measures of climate sensitivity are usually economic, other indices are also available, such as the number of people affected by a climate variation in terms of dietary levels,[13] or in terms of the patterns of land ownership.[14] However, the primary appeal of monetary economic indicators is their ease of intercomparison, and the most extensive effort so far to assess relative economic sensitivities of different sectors and impacts to climate variation is that of the Climate Impact Assessment Program (CIAP)[15] which sought to formulate mathematical relationships between long-term climate change and many economic activities. Results of one calculation are presented in Table VI.2. Other studies of note include those the Oklahoma Climatological Survey[16] who have employed input-output and other economic models to assess the effects of climate variability on a range of sectors and spatial scales.

Table VI.2 Estimates of economic impacts of a hypothetical global climatic change (-1°C change in mean annual temperature, no change in precipitation)

Impact studied	Annualized cost -- 1974 (millions of US dollars)
Corn production (60% of world)	+21
Cotton production (65% of world)	−11
Wheat production (55% of world)	−92
Rice production (85% of world)	−956
Forest production	
(a) US	−661
(b) Canada	−268
(c) USSR (softwood only)	−1383
Douglas fir production (US Pacific Northwest)	−475
Marine resources (world)	−1431
Water resources (2 US river basins)	+2
Health impacts (excluding skin cancer) (world)	−2386
Urban resources (US)	
wages	−3667
residential, commercial	−176 lower bound.
and industrial fossil fuel demand	−232 upper bound.
residential and commercial	
electricity demand	+748
housing, clothing expenditures	−507
public expenditures	−24
aesthetic costs	+219

Source: After Maunder and Ausubel (1985), from Climate Impact Assessment Program (1975).

B. WEATHER AND CLIMATE: THE CHALLENGE OF OPERATIONAL DECISION-MAKING

While it is reasonably easy to make assessments of the general relationship between weather and climate factors and some aspects of production or consumption, the more precise relationships necessary for operational decision-making are much more difficult to formulate. Moreover, even with a perfect weather or climate/economic model, a major problem will be its acceptance by decision-makers. Indeed, it is the acceptance and use of customer-orientated weather and climate information, including commodity-weighted weather and climate information and forecasts of production resulting from this information (as discussed in Chapter X), that offers considerable marketing challenges.

A key question is where do we go from here? First, the impacts of weather and climate on production, consumption, and prices need to be assessed and presented in terms of production figures, costs, or other measures which can be used directly by economists, agriculturalists, planners, and politicians. Second, national meteorological services and private consulting companies must do more to encourage personnel who have a background that will allow them to become 'development' or 'application' and/or 'marketing' meteorologists and climatologists. Indeed, as noted in Chapter II the purely physical and mathematical approach followed by many meteorologists and climatologists often results in their being unaware of the socio-economic applications of meteorology and climatology. A number of national meteorological services including Canada, Japan, New Zealand, France, Sweden, United Kingdom, and the United States have however reacted positively to these comments and it is now recognized that the days of the 'purely physical and mathematical approach of conventional meteorologists' are rapidly becoming a thing of the past. The innovative work of the Canadian Climate Centre in fostering the marketing of climatological information and advice[17] is particularly noteworthy in this respect.

The opportunities provided through education, marketing, and real operational decision-making in the weather and climate business are now key issues in meteorology and climatology. In this regard the 'lead' role of the World Meteorological Organization (WMO) should be highlighted, and in particular the positive response of its Technical Commissions to these issues. For example, at the Ninth Session of the WMO Commission for Climatology held in Geneva in December 1985, it was considered extremely important that application activities be actively encouraged, and that meteorologists and climatologists have to understand problems from the users' point of view. Similarly, the Commission endorsed the viewpoint that for many operational purposes new approaches are needed to tailor short-term weather forecasts and weather information packages to meet the requirements of specific customers, and that both governments and users (including the public) need to be better informed as to the relevance and value of both weather and climate services.

C. THE IMPACT OF CLIMATE

1. An Overview

There are many interconnected components that are involved in climate impact studies (Fig. VI.1). Specifically, society and nature combine to influence societal structures which in turn are related to the perception of impacts, the impacts on specific human and economic activities, and the public demand for action and/or adaptation to climatic variations. A political framework must also be placed over these various interconnections, because the strategies adopted in a centrally planned economy will be different from those adopted in a developing country or a developed market economy. For example, if a climate variation or change is forecast, decision-makers in a developed market economy such as Australia are

Figure VI.1 The interconnected components that are involved in climate impact studies

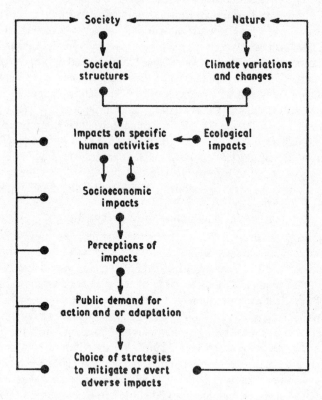

Source : After Maunder (1985), from Kellogg and Schware (1981)

likely to make *independent* decisions on the likely strategies that they could adopt to minimize such impacts. In this case some guidance would come from government agencies, but the individual Australian wheat grower for example would make the final decision as to his best course of action. In contrast, in a centrally planned economy governmental agencies would have a much stronger influence on the strategies that would be adopted.

Although many of the climate/society interactions regarding impacts were first identified and developed during the 1964-70 period,[18] there are still very few case studies which give the dollar impact, or measure the sensitivity in dollar terms, of a specific weather or climate event, particularly at a regional or national level, and much of the thinking on climate sensitivity and climate impact must remain tentative. However, a review of the recent research in this area provides a valuable starting point.

2. Climate Impacts Research: Two Examples

The work of the Oklahoma Climatological Survey[19] is of considerable importance for its strong economic base. In the forward to a series of publications editor Amos Eddy says that use is first made of current economic modelling practice in order to infer the optimal methodology to be employed in analysing the effect of climate on the United States economy, and that this analysis permits significant questions to be answered concerning the sensitivity and responsiveness of the economy to climate fluctuations.

The questions to be asked and the type and scale of the model required to produce the right answers are ultimately linked; however, of prime importance is finding out what kind of questions concerning economic-climate relationships *should* and *can* be asked. Of particular interest is the ranking of such questions with respect to the benefits and costs associated with obtaining answers to them.

A further important consideration in these studies is that the economic impact of the climate includes not simply the passive response of production and consumption to climate variations, but also how climate *information* may be used both to stabilize an economic system and to optimize the benefit/cost ratio with respect to planning strategies and the actions taken by individual sectors. A specific application of using climate information is in food production where significant fluctuations occur from year to year in respect to the total production, and the sub-totals of production in various geographical (and economic) regions. More importantly, 'runs' of several years of 'bad' weather must be expected, and runs of 'bad' weather for food production in one region, but 'good' in another, are also to be expected climatologically. However, since both overproduction and underproduction can cause instability within an economy, the ability to be able to anticipate such occurrences in order to provide for corrective measures is highly desirable.

As an example of this, a central marketing agency can use stockpiles of grain, meat, or wool to keep prices at a particular level, and to increase reserves in times of high production for the same purpose. Clearly, this economic stabilizing activity is in part the result of the weather/climate mix and is a good example of how weather and climate sensitivity not only crosses regional and national boundaries, but also is a critical factor in marketing, political, and strategic decision-making.

A second econoclimatic study is the work of the U.S. Center for Environmental Assessment Services (CEAS)[20] which from 1978 to 1986 made routine weekly, two weekly, and monthly climatic impact assessments of a number of countries. In particular, the reports qualitatively assessed the impact of major climate and other natural events on eight broad categories of United States societal activities. The impact assessments considered those unusual or abnormal meteorological or geophysical events (that is, unusual in time, location, intensity, frequency or persistence) that are (or were) likely to have an impact on societal or economic activities in a special and significant manner. The basis of the studies were the sensitivity of the gross national product to widespread anomalous weather as given earlier in Table VI.1.

An important feature of climate impact studies is that although last month's weather or climate is history, the measurement of its economic impact is not. For example, economic 'indicators' take time to compile and are usually available only several weeks after the period being assessed, while economic 'statistics' take months and in some cases a year or more to compile. However, weather and other environmental satellites, combined with the traditional global weather data collection systems, provide worldwide measurements of the weather within minutes of the 'events' taking place, and *provided* appropriate analyses have been made these data can be converted (*in real-time*) into economic impacts. They can therefore become estimates of that part of the economy that is affected by the weather and the climate. That is, when weather and climate is a limiting factor its impact is known immediately. Thus a forecast of economic data - which at the time the forecast is made have not been compiled, collected, or analysed - can be made using real-time weather and climate data. Moreover, because weather and climate is 'outside' the economic system, its direct effects are easier to measure than the effects of other economic variables.

WEIGHTING THE WEATHER

A. THE VALUE OF WEATHER AND CLIMATE TO AN AREA

Irrespective of the probability of a significant climate warming in most areas of the world in the 21st century as a result of the 'greenhouse effect', by far the most significant weather and climate variations in the coming 20 years will still occur from day to day, week to week, and month to month. To determine the value that can be placed on these short-term weather and climate variations require the identification of those activities that are specifically affected by these variations.

An essential part of any econoclimatic model is the calculation and application of commodity-weighted weather and climate indices for specific areas. These indices are assessed by assessing the significance of various sub-areas to the total 'economic welfare' of a region or a nation (such as retail trade, energy sources, unemployment, or corn production). How is this done? First, the value of weather and climate to areas is determined through the identification of those activities directly or indirectly affected by both the actual weather and climate, *and* by weather and climate information; second, commodity-weighted weather and climate indices are assessed for these activities. In this regard, it must be emphasized that traditional climatological information - produced several weeks after the event - is of limited operational use. Accordingly, to assist significantly in national economic planning, meteorological services needs to provide decision-makers not only with real-time weather and climate information but also information appropriately commodity weighted by areas.

B. COMMODITY-WEIGHTED WEATHER INDICES: THE WHY, WHAT, AND HOW

1. An Overview

To provide appropriate guidance to decision-makers weather and climate information must be *commodity* weighted, for in applying such information, all things and all areas are *not* equal. For example, a forest industry is basically interested in the weather and climate conditions pertaining to the forested areas, and a wheat industry in the weather and climate conditions pertaining to wheat areas. Similarly an electricity industry is interested in the weather pertaining to the populated and industrial areas with respect to electricity consumption, but in other respects to energy source areas and transmission routes.

A key problem associated with the development of such indices is that most commodity-type information refers to large areas of a nation such as counties, states, or provinces whereas weather and climate data are usually related to specific points. Thus, any study which endeavours to associate weather and climate conditions with economic activities such as 'housing starts' in the United States, rice production in Thailand, corn production in Iowa, or energy consumption in Brazil, is faced with the incompatability of the commodity/economic and weather/climate data. The task therefore is to adjust the weather and climate data to fit the available economic data. One way to do this is to use the economic data collated on a county or equivalent basis, and this method has been used in an operational manner by the New Zealand Meteorological Service since the mid-1970's.

Similar data for some other countries is also computed operationally, but except for population-weighted temperatures and heating-degree data for the United States and Canada, it is believed that the comprehensive commodity-weighted aspects of the national indices produced in New Zealand are unique.

2. The New Zealand Experience

The availability of quite complex computer programming techniques has enabled the provision of real-time weather information specifically tailored to national commodity sectors, such as the wool industry, horticulture, and energy production. The potential availability of such information particularly through videotex type systems means that much more useful applied meteorological and climatological information is available on demand. Such information includes indices of weather and climate activity which not only are updated daily (and in some cases hourly), but also are available on a regional and a national basis for a wide variety of commodities.

Specifically, weather and climate indices weighted according to the distribution of 120 economic and agricultural parameters are currently (as of September 1988) analysed by the New Zealand Meteorological Service for several areas and the nation as a whole. The actual weightings used in each analysis are based on the contribution of the 'geographical county' to the total 'population' or 'area'.[1] In 1970 when the original analysis was made, there were 110 counties in New Zealand, and economic data from these countries, together with the weather data applicable to one or more weather stations in each county, were used as the basis for the formulation of the weighted weather indices.

The basis for calculating the weather data weightings is the contribution of the county to the New Zealand total population or area. For example, Waimate County has 0.3% of the New Zealand human population, 1.4% of the land area, 1.9% of the sheep population, 0.1% of the dairy cow population, 0.7% of the beef cattle population, and 5.0% of the crop area. Data showing the relative significance of six counties for 23 economic parameters is shown in Table VII.1 which indicates the relatively large differences in the economic base of various areas of New Zealand.

Table VII.1 Relative significance* of geographical counties in New Zealand

Economic parameter	Geographical county					
	Waikato	Cook	Horowhenua	Mackenzie	Waimate	Vincent
Human population	1.0	1.3	1.0	0.1	0.3	0.3
Sheep	0.8	1.9	0.4	1.3	1.9	1.2
Dairy cows	5.5	0.3	1.8	0.0	0.1	0.1
Beef cattle	1.3	4.1	0.6	0.5	0.7	0.4
Crop area	0.4	0.9	0.5	1.6	5.0	0.8
Orchards	0.6	3.5	0.1	0.0	0.3	9.1
Market gardens	1.3	3.0	4.9	0.0	0.6	0.2
Potatoes	1.0	0.4	7.1	0.1	2.3	0.1
Wheat	0.0	0.1	0.0	1.5	7.3	0.7
Grazing land	0.1	0.2	0.1	9.6	3.0	8.2
Breeding ewes	0.7	1.8	0.4	1.2	2.0	1.2
Sheep/hoggets shorn	0.9	1.9	0.4	1.3	1.9	1.2
Hydro capacity	0.3	0.3	0.0	7.5	1.3	0.5
Apples	0.1	0.5	0.1	0.0	0.5	1.0
Pears	0.1	0.4	0.1	0.0	1.2	1.7
Cabbages	1.3	1.6	7.0	0.0	0.9	0.1
Lettuce	1.5	2.3	6.1	0.0	3.5	0.0
Sweet corn	2.1	49.6	0.3	0.0	0.1	0.0
Outdoor tomatoes	0.8	11.4	2.8	0.0	0.0	0.0
Sheep/beef units	0.9	2.6	0.4	1.1	1.6	1.0
Wool production	0.5	2.0	0.3	1.4	2.1	1.3
Biomass (maize)	3.0	1.2	1.4	1.2	2.1	0.0
Biomass (beefs)	2.4	1.2	1.0	0.7	1.8	0.6

Source: After Maunder (1985).
Note: * Percentage of the New Zealand total.

An example of the computation of part of the national index is given in Table VII.2 for the South Canterbury/North Otago area. It shows that in May 1983 the weighted regional rainfall indices for that area were 91 (% of average) for the land area, 63 (%) for the human population, and 80 (%) for the sheep population. Similar computations were made for all other areas, the weighted indices for all weather stations then being combined to obtain the weighted index for New Zealand.

Table VII.2 Weighted rainfall computations for the South Canterbury/North Otago region of New Zealand for May 1983

Climatological station*	Rainfall % of normal	Weightings					
		Land area		*Human population*		*Sheep population*	
		A	*B*	*A*	*B*	*A*	*B*
Lake Coleridge	129	1.1	142	0.1	13	1.7	204
Ashburton	77	1.2	92	0.8	62	2.4	185
Lake Tekapo	108	2.7	292	0.1	11	1.1	119
Timaru Airport	46	0.8	37	0.4	18	1.1	51
Timaru	48	0.3	14	1.2	58	0.7	34
Waimate	60	1.4	84	0.3	18	1.9	114
Tara Hills	146	1.2	175	0.1	15	0.7	102
Oamaru	54	1.2	65	0.9	49	1.0	54
Region		9.9	901	3.9	244	10.8	863
Weighted Index (B/A)		$\frac{901}{9.9} = 91$		$\frac{244}{3.9} = 63$		$\frac{863}{10.8} = 80$	

Source: After Maunder (1985), updated from Maunder (1972a).
Notes: * These climatological stations are located in and assumed to be associated with various geographical counties. For example, it is assumed that 1.7% of the total New Zealand sheep population is climatologically associated with the Lake Coleridge climatological station.
A= Percentage of the New Zealand total of the economic parameter located in the geographical county and associated climatological station(s).
B= (Column A) x (rainfall at the climatological station expressed as a percentage of normal).

Weighted weather and climate information can be assessed in either real-time or non-real-time. The difference is significant, and relates to the time between when the event occurred, and when information about it is known. Real-time information is now available in New Zealand for combinations of 8 regions, 120 commodities, and 31 climate parameters, the data being available for any sequence of days as well as for the traditional calendar month. A typical example of the data available for the 100-day period ending on 2 July 1984 is shown in Table VII.3. Another example of real-time weighted weather information is given in Table VII.4. The national weather indices given in Table VII.4 are for four key aspects of the New Zealand economy: rainfall in the hydro-electric generation areas, minimum temperatures in the urban areas, rainfall in the agricultural areas, and maximum temperatures in the agricultural areas. This information relates to the previous 7 days, 2 weeks, 4 weeks, 3 months, 6 months, and 12 months to 10 October 1979.

Table VII.3 Typical (unweighted) climatic data for four stations in New Zealand, and weighted climatic data for two regions, and six national commodities (data cover the 100-day period ending on 2 July 1984)

'Weighting'	*Rainfall percentage**	*Temperatures***	
		maximum	*minimum*
Station			
Auckland	68	0.0	+0.1
Napier	31	+0.8	−0.2
Wellington	69	+1.0	+0.9
Alexandra	90	+1.5	+0.4
Area			
Waikato/Bay of Plenty	68	+0.3	+0.4
Canterbury	47	+1.0	0.0
Commodity			
Beef cattle	64	+0.5	+0.3
Sheep	65	+0.6	+0.3
Cows	67	+0.3	+0.3
Wheat	57	+0.8	0.0
Vegetables	43	+0.7	−0.2

Source: After Maunder (1984a).
Notes:
* Percentage of the average for the 100-day period.
** Deviations from the average for the 100-day period.

As previously pointed out such weighted weather information can be assessed in either real-time, or non-real-time. For example, if at say noon today the total rainfall from 50 places in a country for the past 45 days is known, then such information may be considered real-time information. In contrast, 'traditional' climate data for a calendar month are in many countries available only after 'mailed-in-records' have been processed at a central office. In reality of course there is no reason (other than economic) why all weather and climate information should not be available in real-time. Indeed this may soon become the norm as the cost of collecting and processing 'mailed-in-records' becomes too expensive.

3. A United States Example

The methodology used in the above examples can be translated to other areas, and one example for the United States is now examined. Translating the New Zealand station/county model to the United States would mean using weather data from about 3,000 non-real-time stations, and/or data from about 1,500 real-time stations. Neither alternative is possible but meaningful commodity-weighted

Table VII.4 National weighted weather indices - New Zealand

Period to 9 a.m. 10 Oct. 1979	Hydro-elect. rainfall (1)	Urban temperatures (2)	Agriculture rainfall (3)	Agriculture temperatures (4)
Last 7 days	90	−0.5	138	−1.2
Last 2 weeks	153	0.0	142	−0.9
Last 4 weeks	122	−0.1	116	−0.6
Last 3 months	113	+0.3	115	−0.1
Last 6 months	102	+0.3	105	+0.1
Last 12 months	111	+0.3	112	+0.1

Source: After Maunder (1979).
Notes:
(1) Rainfall (weighted by the significance of hydro-electricity generation stations and their catchment areas) expressed as a percentage of the average for the period stated.
(2) Minimum temperatures (weighted by the significance of urban areas) as a difference from the average for the period stated.
(3) Rainfall (weighted by the significance of agricultural areas) expressed as a percentage of the average for the period stated.
(4) Maximum temperatures (weighted by the significance of agricultural areas) expressed as a difference from the average for the period stated.

weather indices can be produced in the United States using real-time or near real-time data from the few hundred weather stations which are available.

As an example, commodity-weighted temperature departures from the average for the United States for the weeks ending 22 October 1978, 2 September 1979, and 25 January 1981 are given in Table VII.5. This shows that within each of these weeks there were very wide differences in the 'response' of different 'commodities' to temperature. For example, in the week ending 25 January 1981 the weighted temperature departures varied from a high +12.7°F (7.1°C) for hogs and pigs, to a small positive departure of +1.6°F (0.9°C) for cotton. These weighted values provide information that is in marked contrast to the traditional 'synoptic' viewpoint in which all points (and areas) are generally considered to be 'meteorologically equal'.

C. NATIONAL COMMODITY-WEIGHTED WEATHER AND CLIMATE INDICES

1. The Setting

As previously noted, a horticultural industry is basically interested in the weather conditions pertaining to horticultural areas, a wool industry in the weather

Table VII.5 United States weighted temperature departures(°F)

Commodity	Weeks ending		
	22 Oct 1978	2 Sept 1979	25 Jan 1981
Employees in transport/public utilities	−1.2	+3.1	+5.0
Electric energy: production	−1.2	+2.7	+4.6
Electric energy: installed capacity	−1.2	+2.8	+4.6
Electric energy: residential sales	−1.3	+2.8	+4.5
Gas: residential sales	−0.6	+3.4	+6.5
Gas: commercial sales	−0.3	+3.1	+6.7
Gas: industrial sales	+0.4	+2.1	+4.7
Interstate highway mileage	−0.3	+2.6	+6.2
Farm products sold	+0.2	+2.6	+8.9
Corn: value	−0.5	+3.5	+11.4
Wheat: value	+1.6	+2.7	+12.5
Cotton: value	+1.4	+0.8	+1.6
Soybeans: value	−0.7	+2.9	+9.7
Cattle production	+0.5	+2.4	+8.8
Hogs and pigs: population	−0.3	+3.3	+12.7
Broilers production	−2.2	+2.0	+1.1
National forests – acreage	+3.1	+1.5	+9.9
New housing units: value	−0.1	+2.4	+5.3
Retail trade: sales	−1.2	+3.2	+5.0
Federal aid (Medicaid)	−1.6	+3.5	+4.9
Coal production	−2.7	+3.3	+4.6
Human population (black)	−2.2	+3.0	+2.6
Human population (white)	−1.2	+3.2	+5.1
Households	−1.2	+3.1	+4.9

Source: After Maunder (1985).

conditions pertaining to sheep areas, and an electricity industry in the weather pertaining to the (human) population areas with respect to consumption, but to energy source areas and transmission routes in other respects. Using such information in appropriate analyses it is possible to 'weather adjust', or at least 'environmentally monitor', indices of national economic activities on a nationwide scale. It is also potentially possible to use such information to provide a forecast of the tendency of national economic indicators before the actual indicators are available, since weather information is available in real-time, whereas national economic indicators usually have a time publication delay of at least two or three weeks - even in the most developed societies.

2. National Weather, Consumption, and Production

Variations in any nation's weather and climate in the recent past cannot of course explain more than a small part of the causes of consumption and production variations. Nevertheless, like the Dow-Jones index (which despite its limited base is considered by many investors to be the leading indicator of stock market performance), it is believed that commodity-weighted weather and climate indices, on a national scale, can provide some guidance to understanding the vagaries of a nation's 'economic climate'. Computer tabulations and graphical analyses of this type of information are produced by the New Zealand Meteorological Service for many time scales and for over 120 commodities and have many applications, such as those discussed in Chapters VIII, IX, and X for the dairy, meat, wool, electricity, retail, and transport industries.

It is considered that the weighted weather and climate indices for sequences of days (such as those shown in Table VII.4) provide a real alternative to the traditional method of portraying weather and climate information which treats all areas as being equal. Rather, what is required is the knowledge that during the last 'n' days, temperatures, rainfalls, soil temperatures, or sunshine - weighted according to the distribution of a specific population such as people, animals, crops, or their productivity - were above or below normal.

3. New Zealand National Rainfall Indices

National weighted rainfall indices show very clearly the variations which occur if different economic parameters are considered. For example, the indices for New Zealand for March 1969 indicated that the relatively dry period which was prevalent at that time was most severe in the dairying areas (index of 27% of average), and least severe in the crop areas (62%).

Commodity-weighted rainfall indices for New Zealand for 'the national dairy farm', the 'national sheep farm', and 118 other 'commodities', including people, energy, and transport, have been computed for each month from January 1950. Examples of the weighted data are given in Fig. VII.1 for the period January 1982 to December 1984.

4. Weighted National Agricultural Soil Water Deficit Indices

The use of traditional monthly climate totals, particularly those related to rainfall, usually assumes that there is an even distribution throughout that period. Daily climate information can be used to overcome this deficiency by using the number of days in *any* period in which there is a 'soil moisture' or 'soil water' deficit (a 'day of soil water deficit' can be defined as occurring when the combined precipitation and soil moisture, assuming a maximum soil moisture capacity of 75 mm, is less than an assessment of the water need).

Figure VII.1 New Zealand rainfall indices*: 1982-84

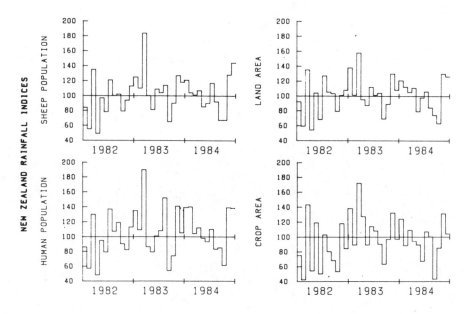

Source: Compiled from data supplied by the New Zealand Meteorological Service.
Note: * Weighted according to various economic parameters.

The weighted number of 'days of soil water deficit' for New Zealand for the 15 seasons 1970/71 to 1984/85 using three different pastoral weightings is shown in Table VII.6. These data for the June-May season and the autumn period can be said to represent the 'national sheep farm', the 'national dairy farm', and the ' national beef cattle farm' respectively. The indices show just how variable is the New Zealand climate; noteworthy, are the very high values in the drought seasons of 1972/73, 1977/78, and 1982/83.

5. Application of National Commodity -Weighted Weather and Climate Indices

The development of commodity-weighted weather indices is primarily aimed at studies which relate weather and climate events to national economic activities. Such research requires in most cases economic data for short time periods, of which there is an almost complete lack of in most countries. Two important series are, however, available for New Zealand and provide useful application examples.

The first concerns the consumption of electric power. Analyses have indicated that a 1°C deviation of the temperature from the average for a particular period is associated with a 2-3% change in the demand for electricity. In a recent analysis[2]

Table VII.6 Weighted* number of days of soil water deficit in New Zealand

Season	June–May			March–May		
	Sheep	Cows	Cattle	Sheep	Cows	Cattle
1970/71	38	16	26	13	3	9
1971/72	31	19	24	6	2	3
1972/73	59	45	52	17	13	15
1973/74	40	35	37	8	4	7
1974/75	30	17	25	6	4	6
1975/76	36	20	25	22	16	17
1976/77	27	24	22	17	15	14
1977/78	55	41	51	23	24	24
1978/79	27	20	28	4	1	2
1979/80	11	3	8	2	0	1
1980/81	38	21	28	12	10	11
1981/82	38	21	30	9	3	5
1982/83	44	40	50	15	13	18
1983/84	20	7	15	6	2	3
1984/85	43	11	31	21	6	13

Source: Updated from Maunder (1980b).
Note: * Weighted by the distribution of sheep, dairy cows, and beef cattle as shown.

a population-weighted temperature index was computed for each month for a number of years. These indices were then correlated with the pattern of electricity consumption. After an appropriate statistical allowance for the 'normal' climate, seasonality, trend, and 'workdays' the results indicated that many of the variations in electric power consumption could be accounted for by the deviation of the weighted temperatures from the average for each period. Such an analysis provided a valuable insight into the weather sensitivity of a key sector of New Zealand's economy; this is important since expensive thermal power stations are used in a back-up role during periods of cold temperatures and hence peak demand.

The second series concerns the important dairy industry which is particularly weather-sensitive. Variations in the day-to-day, week-to-week, and month-to-month production are of specific interest to the marketing and transport sectors. Further details of the dairy production/weather models are given in Chapter X.

THE AGRICULTURAL SCENE

A. DROUGHTS

1. Drought Impact Pathways

The human consequences of a severe drought depend largely on the ways in which the effects filter through the socio-economic and political fabric of a society. For example, reduced agricultural production or changes in hydro-electric output are not the *ultimate* concerns, but rather the degree to which people are affected in terms of income, health, and community stability. That is, a reduced wheat yield does not - except in certain critical circumstances - directly affect a society, but a change in the price for wheat or an inferior substitute does.

A framework for tracing the impacts of drought occurrence in the Great Plains of the United States[1] is shown in Fig. VIII.1. It depicts the pathways that drought

Figure VIII.1 Hypothetical pathways of drought impacts on society

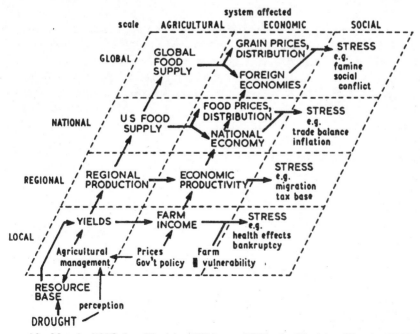

Source: After Maunder (1985), from Warrick and Riebsame (1981), after Warrick and Bowden (1981).

impacts could take spanning local through to global spatial scales, and three major systems that could be affected (agricultural, economic, social). As indicated, the initial 'disturbance' originates from a meteorological event, and becomes an agricultural drought - as distinct from a meteorological drought - when agricultural production falls below a perceived threshold. The agricultural drought then translates into a drought impact when the stress is detected in the economic, social, and political sectors. The degree to which the initial climate event is transformed into stress is influenced by a number of factors including market prices, government policies, farm stability, and the degree to which the drought is seen as a local, regional, or global problem.

2. Costs of Drought: Problems of Measurement

The costs of droughts are difficult to estimate, and over two decades ago an issue of the Australian *Current Affairs Bulletin* on 'Drought' rightly pointed out that it is often very difficult to establish real cause-and-effect relationships. In addition, there are - as they said[2] - other 'book-keeping problems'.

> The mind may well boggle at estimates of the cost of a drought calculated to include wool shorn from sheep never conceived and wheat not harvested from crops never sown. Even if it can be admitted that one can in some way lose something which never came into being, the calculation of values of these things becomes very involved.

A critical insight into the problems relating to the evaluation of drought conditions has been made by Heathcote[3] who indicated that there were four main effects: the abnormal service capacities to meet abnormal demands for water, such as storage well in excess of normal use; the maintenance of a foreign credit reserve to cover the fluctuations of national production; the cost of drought research; and the reticent attitude to resource development in the drought-prone areas. He also pointed out that the 'negative' aspects of droughts include the spasmodic effects which are relatively short run, and the incessant effects which result from the costs of preparing against the recurrence of drought.

A more specific study of a severe drought in New Zealand by the author[4] attempted to measure in dollar terms the impact of the drought. Calculations suggested that total costs exceeded 3% of the total value of exports from New Zealand.

3. Beneficial Effects of Drought

The effects of drought are not always negative. Indeed in certain ecological contexts it has been shown that the effects are sometimes beneficial. For example,

Perry[5] found that the drought of 1958-61 in Central Australia had caused little long-term damage to the vegetation, and that the perennial grasses were in a healthier condition than would have been the case after a 'normal' succession of seasons because livestock deaths had reduced the pressure on grazing.

In addition there are the more direct economic benefits, for to many people drought clearly means increased business through the cartage of animals, water, and feed supplies. Moreover, in many cases those receiving the benefits are in a region or a country that is not affected by the drought. For example, a drought in Europe's dairying area brings substantial benefits to the dairy farmer of New Zealand some 20,000 km away, in the form of additional export opportunities. Drought may also have much wider international implications than the simple export or non-export of a commodity.

4. Drought and an Intense Freeze: An Example

If we are to use the 'climatic income' which is available to various sectors including agriculture, we must learn how to cope better with the wide swings of nature's pendulum. One specific and dramatic swing of the pendulum occurred in the winter of 1976/77 in the United States when a severe drought in the west was associated with an intense freeze in the east. The direct losses of this climatic swing were estimated[6] to be $US27 billion (in 1977 dollars), including $9 billion in the production sector, $7 billion in the foodstuff sector, and $5 billion in the transportation sector.

The impact was most severe during the six-week period from January through to mid-February (see Fig. VIII.2). In assessing such losses, it is important to appreciate that *nationally* it is the net loss (or net cost) that is important, irrespective of the fact that some sectors, including and excluding agriculture on both regional and national scales, may in fact 'gain' from the occurrence of severe weather. In addition an important question is what proportion of these losses could have been averted through the optimal use of weather information and forecasts.

5. Drought in Canterbury, New Zealand: What the Climate Record Revealed

A severe drought in the Canterbury area of New Zealand occurred in 1981/82. The following abridged account was originally written by the author[7] in September 1982 *while the drought was happening* and provides a useful example of the day-to-day information that was available, and the decision-making that was necessary at that time. It is typical of the kind of information that would be available in many other drought areas, and it shows that decisions are always more difficult to make 'before the rains come' than after the grass or crops have become green again.

The drought in Canterbury was one of a large number of droughts that have affected this area since early times. In purely economic terms, droughts are often

Figure VIII.2 Costs of the 1976/77 intense freeze and drought in the United States

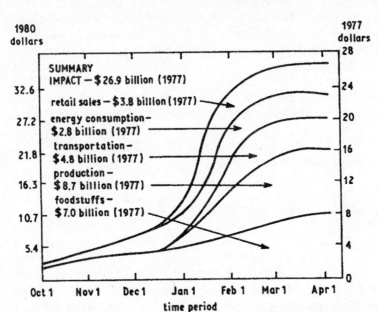

Source: After Maunder (1985), after Center for Environmental Assessment Services (1982).

difficult to classify or rank, but whatever the kind of drought, there remains the problem of 'too little water' required by 'too many things'. This 'too little water' concept was all too readily apparent in Canterbury at the date of the original analysis (early September 1982).

At Christchurch, a typical station in the drought area, the total rainfall for the 10 months November 1981 to August 1982 was 308 mm or 54% of the average for this period. Comparable records at this site date back to the 1890s, and the 1981/82 total was the lowest ever recorded, the previous lowest November to August total of 314 mm occurring in 1896/97. Data for the driest November to August periods are given in Table VIII.1. The table also shows that the rankings of the lowest November to October rainfalls are similar to the rankings of the November to August rainfalls.

In terms of soil moisture the 1981/82 season was also extremely dry with 79 'days of soil water deficit', the highest value for this period since comparable computations were started in the 1949/50 season. If the 68 days of soil moisture deficit in the previous 1980/81 season are added to this total, then for the two successive seasons (1980/81, 1981/82) the combined total was 147 days. This may be compared with the previous highest two successive season totals of 133 days in 1971/72/73, 128 days in 1970/71/72, and 106 days in 1954/55/56.

Table VIII.1 Christchurch 'seasonal' rainfall rankings: 1896-1982

Season*	November to August		Season*	November to October	
1981/82	308**(54)***				
1896/97	314	(56)	1896/97	397	(59)
1931/32	334	(59)	1968/69	404	(61)
1968/69	337	(60)	1921/22	423	(64)
1921/22	347	(61)	1914/15	439	(67)
1970/71	362	(64)	1932/33	442	(67)
1932/33	364	(64)	1970/71	443	(67)
1947/48	380	(67)	1947/48	449	(68)

Source: After Maunder (1982b).
Notes:
  * Ranked in order of rainfall totals.
  ** mm.
  *** Percentage of the 1941-70 normal rainfall.

To the concerned farmer and community in which he lives, two questions arise. First, does long-range weather forecasting provide any answers; second, is weather modification in the form of rain-making possible? Regrettably the answer to both questions is no, for although much work has been done on both subjects, there are few positive results which could provide any real help to the communities affected by this drought. The most useful guidance that could be given therefore is that, in terms of historical droughts, the current drought (at least in a climatological sense) is (or was) among the most severe. In addition, the historical record indicated that no immediate change in the situation was likely, and that from a planning viewpoint, the best decision was to assume that the drought will continue. (In retrospect, the *actual* rainfall at Christchurch for the September 1982 to March 1983 period was 301 mm or 86% of average, while the *actual* November 1981 to October 1982 rainfall was 410 mm or 62% of average, making this period the third lowest on record.)

6. Drought in Africa: The Sahel Experience

'Drought. Africa. These two words have been linked together in newspaper headlines around the world for more than a decade.' These words in the April 1985 issue of *Weatherwise* are used to introduce a report by Winstanley[8] on 'Africa in Drought : A Change of Climate?'. The paper provides a very useful survey of the African drought situation, and Winstanley comments:

Tropical Africa has suffered a decade or more of recurring droughts. This has been an important cause - but not the only cause - of famines, economic problems, and social disruptions. ... (the) close association between rainfall and development raises many questions that have important strategic and policy implications, not only for the peoples of Africa but also for the international community.

Winstanley then goes on to discuss the climatological aspects of the African drought(s), and in a special section highlighting the climate impacts of the drought(s) he asks two questions: what are the effects of repeated and persistent droughts in the sub-Sahara zone of the last decade, and what are the possible effects of a continuation in the trend toward lower rainfall? Possible answers in relationship to the impacts of the drought(s) on rain-fed food production, cash crops, inland fisheries, river basin development, economic development, and desertification are discussed.

Clearly drought as severe as has occurred in Africa since the late 1960's calls for action, and Tooze[9] writing in *Nature* (9 February 1984) stated that conditions in the Sahel are part of the climatic threat to the world's poorest people. It was noted that according to the Club du Sahel, aid to the area has doubled in ten years and now amounts to 15% of the total gross national product. Tooze then points to the serious problems associated with aid assistance to areas suffering drought and comments: '... so dependent on aid has the region become that some agencies now fear that long-term massive food aid is sapping the ability of the region to acquire the resources necessary to respond to normal climatic swings, let alone to cope with disaster.'

Reference should also be made to the booklet *Climate Variations, Drought and Desertification* written by Hare[10] and published by the World Meteorological Organization in 1985. Hare comments that around the world's deserts there is a wide extent of semi-arid and sub-humid land supporting a large human population, and that in spite of its hazards, the arid zone has offered a challenge to humanity that for centuries has been successfully answered. He observes however that today this zone is in many places the scene of acute distress and even of tragic famine, and that in spite of the long traditions of human adaptation, many nations now find themselves unable to wrest a reasonable living from the vanishing soils and natural resources.

Four questions are posed by Hare in relation to the above: what accounts for this unhappy situation, is it human failure, are the causes to be found in human interference, or is it a natural deterioration of climate? The answers to these questions, according to Hare, involve the problem of desertification - the spreading of deserts into formerly productive land, or more correctly the degradation of land until it can no longer adequately support living communities - and the (simple) concept that drought occurs when there is too little water for too many people, animals, or crops. Hare also states that '... climate alone will not destabilise the productive cloak or life that shelters the world and yields so much to human needs

- but, nevertheless, it cannot be disregarded'. Hare's comments are particularly relevant and highlight the fact that the spectrum of drought in Africa involves both risks and opportunities, and people and nations ignore the climate factor at their peril.

B. PASTORAL PRODUCTION AND WEATHER RELATIONSHIPS

1. The World Scene from a New Zealand Viewpoint

As an example of the sensitivity to weather and climate of pastoral production, New Zealand has a unique situation. Indeed it is considered by many observers that New Zealand has a significant climate advantage for grassland agriculture, this advantage being translated into the development of New Zealand into the world's leading exporter of grassland products (Table VIII.2).

Table VIII.2 New Zealand and world production and food exports compared*

Product	New Zealand exports as percentage of world exports	New Zealand exports as percentage of New Zealand production	World exports as percentage of world production	New Zealand production as percentage of world production
Lamb and mutton	50.0	80.9	14.4	8.9
Beef and veal	8.9	46.7	6.3	1.2
Butter	20.8	76.8	15.0	4.1
Cheese	6.6	97.5	11.7	0.8
Dried milk	10.6	85.4	39.5	4.9

Source: After Maunder (1980a), data from New Zealand Monthly Abstracts of Statistics, FAO Production/Trade Yearbooks. Original table was published in *New Zealand in the Future World*, a booklet on 'sustainability' prepared by Diane Hunt for the NZ Commission for the Future, Government Printer, Wellington, 1979.
Note: * Based on 1977 production.

In terms of world wool production (as just one example[11] of pastoral production) it is useful to compare the 'New Zealand wool climate' with the climates of the other main wool production areas of the world, and to assess what climate advantage New Zealand has over her competitors. (A similar comparison could be equally well presented comparing say the Thailand climate with the climate of the other chief rice production areas of the world and then to assess the climatic advantages which Thailand has over her competitors.) In this example, climate data for representative stations in each of the major wool producing countries of the world were analysed and compared with relevant data for New Zealand stations. Only temperature and rainfall data are readily available for most wool producing areas, but as a first approximation such data may be translated into 'pasture growth/wool production potential'. A summary for the wool production areas in the form of the number of 'moist' months and 'growing' months is given in Table VIII.3

Table VIII.3 Wool products - selected world climate data

Stations	'Moist' months*	'Growing' months**
New Zealand		
Whatawhata	12	12
Omarama	11	8
Mid Dome	12	8
Argentina		
Buenos Aires	12	12
Bahia Blanca	8	12
Santa Cruz	0	7
Australia		
Hay	1	12
Dubbo	12	12
Bourke	1	12
Port Augusta	0	12
China		
Hanchow	3	7
T'ai-Yuan	4	7
South Africa		
Grootfontein	5	12
Victoria West	2	12
USA		
Abilene	7	12
Austin	12	12
Cheyenne	5	6
USSR		
Ordzhonkikdze	7	6
Astrakhan	0	7

Source: After Maunder (1977c) from Maunder (1963).

Notes: * Number of months with average rainfall 38mm or more (and index of 'moist' months).

     ** Number of months with average daily maximum temperature of 13°C or more (and index of 'growing' months).

2. Livestock Investment Response: A New Zealand Example

Despite the 'well-watered' image that New Zealand has there are quite significant climate fluctuations from season to season. Most agrometeorological research has traditionally focused attention on activities such as marketing, production, animal health, crop protection, and crop quality. An additional factor is the farmers' investment response. This important aspect has been examined by the Reserve Bank of New Zealand[12] which considered that the main factors influencing farmers' livestock investment response in New Zealand are monetary exchange rates, climate conditions, biological factors, technological change and innovation, and the availability of land, labour, and credit.

The Reserve Bank in its report considered that climate conditions provide a major constraint to growth, since livestock feed is largely grown on the farm and it is not usually feasible or economic to import large quantities of supplementary feed. The effect of climate as a constraint on livestock investment by farmers is therefore largely related to pasture production and animal carrying capacity, especially during the winter period. The effect on reproductive rates is also significant since feed intake prior to conception and birth has a strong positive influence on lambing and calving percentages. Drought has an obvious direct effect on carrying capacity and performance, but it can also have a significant lagged effect through to the following season.

Data for the 1960/61 to 1979/80 period for sheep and beef productivity in New Zealand measured in terms of stock units, the monetary exchange index, real expenditure per stock unit, and the climate variable 'days of soil water deficit', were examined in the study. The year by year changes give some indication of the nature of livestock investment responses, but a clearer picture can be obtained by aggregating the data into periods which have similar characteristics. It is convenient to look at three five-year periods 1962/63 - 1966/67, 1967/68 - 1971/72, and 1972/73 - 1976/77 as shown in Table VIII.4. In commenting on the results, the Reserve Bank said that the overall weather and climate conditions in New Zealand during the 1970-79 decade were not conducive to creating confidence in the farming community. For example, there were very high numbers of days of soil water deficit in the seasons 1972/73, 1973/74, and 1977/78. The impact of these droughts, coupled with the rapidly increasing costs of farming, both on and off the farm, need to be much more appreciated if livestock farming in New Zealand is to continue to prosper.

C. THE COMMODITIES MARKET

1.Weather and the Futures Market: A United States Example

One of the most significant weather-related subjects is the use of weather and climate information in the 'futures market' in which large financial gains (and

Table VIII.4 Agro-economic and climatic factors: the New Zealand sheep/beef farm - selected periods

Annual average	Period 1 1962/63– 1966/67	Period 2 1967/68– 1971/72	Period 3 1972/73– 1976/77
Sheep and beef stock units (Average annual % change)	+4.1	+0.4	+1.4
Terms of exchange index	1049	954	1074
Real expenditure per stock unit (1975/76 $)	$8.87	$8.24	$9.17
Weighted no. of days of soil water deficit	28	32	37

Source: After Maunder (1986b), from Walsh (1981).

losses) can occur. In contrast, few investors in the share market would consider that there is any similar link between weather variations and share prices, although the more astute investor realizes that droughts and floods often have specific effects on the prices one has to pay for some shares, or the prices one is able to realise when those shares are sold. In the futures market, however, prices clearly vary on the basis of weather information.

The futures market is the exchange where commodities such as wheat, corn, soybeans, sugar, wool, cocoa, and frozen orange juice are bought and sold for delivery at a stated time in the future. The market exists to enable people who actually trade in commodities such as farmers, chocolate manufacturers, and cooking oil wholesalers, to reduce the risk they face of losing money because of changes in the prices of specific commodities; to allow producers to get a guaranteed price for raw materials they haven't yet produced; and to allow manufacturers to know exactly how much they will have to pay for raw materials they will need in the future. The prices for commodity 'futures' are related to a number of things, the most important being the anticipated demand for and supply of a commodity at some future date. Consequently, there is a close relationship between the actual, reported, and expected weather conditions in the various producing areas, and the price at which 'futures' in a commodity will be bought and sold. Reports of *past* weather and forecasts of the *expected* weather therefore have a real dollar value as far as the commodity markets are concerned, and clearly affect the prices that the consumer will eventually pay for these commodities.

The commodities market is perhaps the best example of weather and climate sensitivity in the business world, as it reflects the sensitivity of *both* the actual and the reported weather on prices. Indeed, reports of frosts in Brazil, freeze conditions in Florida, or droughts in Australia, can affect commodity prices *before* the actual weather conditions are known. But it must be emphasized that although official weather and climate data are collected, analysed, assessed, and reported in real-time, the data are at best sample surveys; moreover, media reports of weather conditions from 'secondary networks' that are not part of the 'official' record are also part of the information package available to commodity market dealers and their customers, and *all* information is acted upon according to the 'sensitivity' of the markets. In addition to the weather that has or is reported to have occurred, is the weather and the climate that is expected to occur in the future. The futures market is therefore a 'melting pot' of past, present, and future weather and climate information, in which at least in one sense the ultimate economic sensitivity of weather and climate is measured.

But despite the sensitivity of the futures market to weather information the current emphasis is surprisingly on providing to the market only basic weather information, and not specific commodity-weighted weather information. Consequently there is a large untapped market for the more imaginative meteorological consultant to assist those people who want to know *much more* about whether to buy or sell their wheat, wool, coffee, or soybeans, and to whom the typical media comment[13] that 'Grain and soybean prices in the United States plunged on Wednesday because it rained in Iowa' is of little *real* value.

2. Commodity Prices: A Specific Example

The impact of a specific climate event on commodity prices can be quite considerable, affecting export markets, the ability to import, and the consumer price index. For example, in the United States the price of soybeans varied (in 1979) from $6.50/bushel in January to a peak of about $7.40 in mid-summer, before decreasing to about $6.30 in December 1979 and $5.90 in May 1980, and then - mainly due to the anticipated weather related shortfall in production - increased very rapidly to over $8.00 in November 1980[14](see Fig. VIII.3). The economic and social implications of such weather-induced prices are many and, coupled with similar price increases in other key food commodities, it is evident that significant weather and climate variations can play a major role in overall price structures both nationally and internationally.

D. CLIMATE PRODUCTIVITY INDICES: NATIONAL AND INTERNATIONAL

1. National Climate Productivity Indices: A New Zealand Example

In many countries the weather and climate are taken for granted, and other factors are advanced to explain the vagaries of a nation's 'economic climate'. But while this 'media and political term' may well be used correctly if only one meaning of

Figure VIII.3 Changes in the US price of soybeans: 1979 and 1980

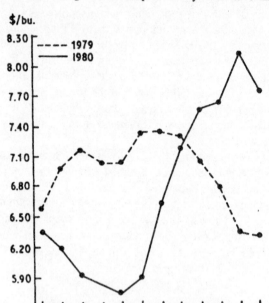

Source: After Maunder (1985), from Liebhardt (1981), data from Center for Environmental Assessment Services (1981).

climate is taken, the atmospheric component of the economic climate must also be considered.

In the belief that real-time information on the agricultural climate of New Zealand is not only important but also essential if the business and farming communities in New Zealand are to maximise their agricultural potential, and at the same time minimize their costs, an 'Agroclimatic Productivity Index' has been devised. The index compares the past and the present weather to give a single figure indicator of what could be called the 'atmosphere economic climate' of the 'national farm'.

Two aspects are considered in computing the 'Agroclimatic Productivity Index': first, 'days of soil water deficit' (or more specifically 'days below wilting point'); and second, the number of growing degree-days (base 5°C). Each factor is in turn weighted according to the significance of the months being considered. For example, the influence of days of soil water deficit is much more significant in mid-summer than in mid-winter. The weighted values are further adjusted by the number of months being considered in the composite index. The final index provides a 'one value' indicator of the overall state of the New Zealand agroclimatological scene, and hopefully the 'New Zealand Farm'.

The New Zealand 'Agroclimatic Productivity Indices' for the consecutive months from June 1977 to May 1985 are given in Fig. VIII.4, each index being a cumulative weighted value of the previous six months. During the eight seasons shown, the index varied from 815 for the six months ending March 1978 (the peak of a drought) to 1,200 for the six months ending February 1980.

Figure VIII.4 New Zealand agroclimatic productivity indices

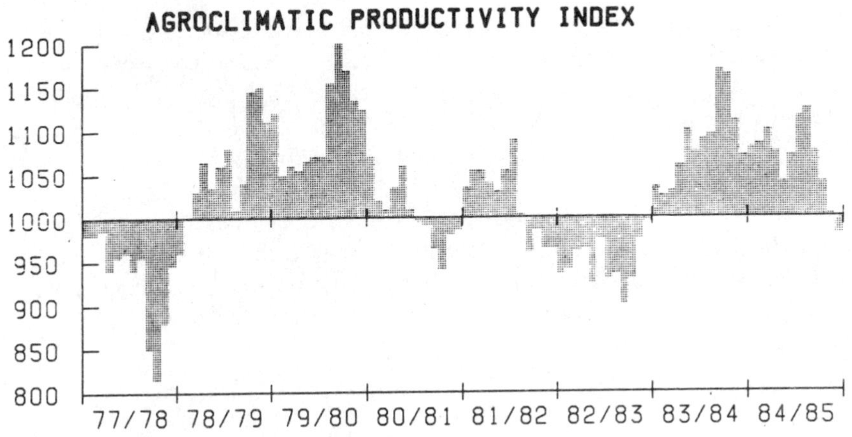

Source: Complied from data supplied by the New Zealand Meteorological Service.

2. International Climate / Agriculture Linkages

The international trade of grains, fruit, and pastoral products are key elements in the economic structure of many countries. Indeed the balance or imbalance between exports and imports of agricultural commodities constitutes an important cause of political and social concern in the world today. Clearly weather and climate conditions in the exporting countries are paramount in these considerations, as are the weather and climate conditions in the importing countries, in that 'good' agricultural weather in the importing country may reduce the need to import certain commodities. Similarly, variations in the weather and climate of the major exporting countries clearly have an affect on the price of those exports.

The ability of all nations to anticipate their agricultural production, and hence their capacity to export and their need to import is important. It is also evident that the economic, social, political, and strategic advantages that weather-based forecasts of such information could give some countries are wide ranging, and involve decision-makers in the production, processing, marketing, and transportation sectors. In the case of New Zealand, her success as an exporter of pastoral products has been highly dependent upon her ability to produce high quality products at relatively low cost. However the situation is especially vulnerable to the costs 'beyond the farm gate'. In this regard the impact of weather variations on agricultural production both within and outside New Zealand is very significant, and in examining the place of New Zealand in the international commodity scene, two examples of the potential value of weather-based pastoral production forecasts are given.

Hypothetical pastoral commodity information based on weather information to a specific date is given in Table VIII.5. The predicted values for the commodities are given as 110 (i.e. 10% above last season) for New Zealand, and 90 (i.e. 10% below last season) for the rest of the world. Using the weightings shown (in this case national exports or production as a percentage of world exports or production) the world pastoral/weather index would have varied from 90 for merino wool and 91 for cheese, to 100 for lamb and mutton, and 107 for cross-bred wool. Such information may be compared directly with the hypothetical New Zealand index of 110. The comparative advantages and/or disadvantages to a country - in this example New Zealand - of a specific season may then be assessed. In a real situation, weather-based forecasts of both national and international commodities including the important crops of corn, wheat, and rice would be progressively updated and monitored.

Table VIII.5 Weighted international pastoral weather indices †

| Commodity | New Zealand | | Rest of World | | World |
	Weight	Index	Weight	Index	Index
Lamb and mutton	50*	110	50***	90	100
Beef and veal	9*	110	91***	90	92
Butter	21*	110	79***	90	94
Cheese	7*	110	93***	90	91
Dried milk	11*	110	89***	90	92
Wool					
Merino	2**	110	98****	90	90
Cross bred	87**	110	13****	90	107
Other	11**	110	89****	90	92

Source: After Maunder (1979).
Notes:
  † Hypothetical (based on weather information up to . . .)
  * NZ exports as percentage of world exports.
  ** NZ production as percentage of world production.
  *** Rest of World exports as percentage of world exports.
  **** Rest of World production as percentage of world production.

The implications of the hypothetical information shown in Table VIII.5 to decision-makers such as the New Zealand Wool Board, Minister of Agriculture, Reserve Bank, Treasury, wool brokers, and the wool grower are potentially many, but clearly the volume of the agricultural commodity (in this case wool) is only *one* factor with which a buyer of wool has to be concerned. For example, the relative value of currencies is clearly reflected in the buying price of most agricultural products. Nevertheless, the key factor in most aspects of agricultural production is the actual physical volume available to market, and as this is clearly weather related, it should be taken into account.

3. International Climatic Productivity Indices

The national 'Agroclimatic Productivity Indices' previously discussed can also be developed for use in an operational manner for specific application to international commodities. Thus 'Agroclimatic Productivity Indices' can be developed for nations as a whole, for commodities for selected nations, and for commodities for combinations of nations. Clearly, a global (or at least a hemispheric) 'Agroclimatic Productivity Index' for corn, wheat, soybeans, wool, meat, coffee, etc., has obvious political and trading overtones.

This further development of national weighted climatic indices was initially developed by McQuigg Consultants Inc.[15] using the technique already described for assessing the 'Agroclimatic Productivity Index' of New Zealand. Table VIII.6 shows a few climate indices for various national commodities for May 1981; these could easily be developed into hemispheric commodity indices.

Table VIII.6 International commodity-weighted agroclimatic indices: May 1981

		Rainfall	Temperature
		(percentage probability)*	
USA	Corn	72	17
	Soybeans	84	2
	Winter wheat	96	49
	Spring wheat	23	56
CANADA	Wheat	17	82
USSR	Winter wheat	36	8
	Spring wheat	28	16
	Sunflower seed	16	4
	Corn	24	4

Source: After Maunder (1984b), from McQuigg Consultants Inc. (1981).
Notes: * 100 = very wet/very warm.
0 = very dry/very cold.

In terms of the impact of weather and climate on a nation, international trade, commodity prices, and food shortages and surpluses, 'agroclimatic productivity' type indices such as those described could provide decision-makers with a good 'one index indicator' of the 'agro-weather-economic' scene. Further, the availability of real-time weather information from around the world, and the availability of reasonably credible 'agro-weather-economic' models, means that a world and/or hemispheric 'weather productivity' index is a reality.

At the international level, such commodity-weighted weather information can be used by one country to monitor and in some cases predict the agricultural production of another country. The economic, strategic, and political 'advantages' that such information could give a country are wide ranging, and involve decision-makers at the government, producer board, producer, and processing levels. The marketing advantages of knowing that domestic production of a commodity is doing comparatively better than that of one's competitor is a challenging prospect, but with the important corollary that astute competitors would also know when one's own country's production was likely to be less than expected.

Chapter IX

WEATHER RELATIONSHIPS

A. ELECTRIC POWER / WEATHER RELATIONSHIPS

1. An Overview of Early Studies

Several studies have been made of the linkages between weather conditions and electricity consumption. In an early paper, Davies[1] commented on the difficulty of making accurate load estimates because the demand is very sensitive to changes in the weather. At around 0°C for example, he noted that in Britain the increase in demand per 1°C fall of temperature was at the time of the investigation in 1960 as much as 290 megawatts (this is equivalent to about 10% of the peak winter load in New Zealand in 1985) and at lower temperatures this rose to 400 megawatts. Other weather elements such as wind, cloud, fog, and precipitation also cause considerable variations in electricity demand. For example, the study showed that near 0°C, a 25 knot (46 km/h) wind increased the demand in Britain by about 700 megawatts as compared with that on a calm day. Cloud also has an appreciable effect on the lighting load, and (at the time of the study) caused a variation in Britain of more than 1,200 megawatts. In particular, dark clouds over London caused an increase in demand of 350 megawatts.

Several additional studies have been published[2] and they emphasize the importance of low temperatures, strong winds, and 'heavy' cloud in increasing electricity consumption. An increasing *summer* demand for power is however occurring in those parts of the world where air-conditioning is used. An interesting side-effect of this development was explored in a study[3] in 1969, which considered the potential decrease of summer daytime temperatures in the mid-west of the United States through the creation of aircraft contrails.

2. Electric Power Consumption and Temperatures: A New Zealand Example

The natural resources of water and geothermal steam are harnessed to provide a substantial proportion of New Zealand's electricity requirements, but an increasingly important source is obtained by coal, oil, and gas fired 'thermal' stations. The mix of non-hydro and non-geothermal electricity generated is particularly important because the cost of producing electricity at 'thermal' stations is several times the cost of producing electricity at the hydro and geothermal stations.

The day-to-day demand for electricity in New Zealand is primarily related to temperature, and discussions with the supply authorities have indicated the value of being able to predict the demand for electricity up to 30 hours in advance. Such predictions allow the phasing of the more expensive 'thermal' generation stations into the New Zealand electricity system in the most economical manner. From an economic point of view, it is therefore very useful to be able to predict the weather conditions likely to cause an increase in electricity demand, and hence the proportion of the national electricity capacity that will be needed to be 'thermally' generated.

The consumption of electricity for each half-hour is available for specific areas in New Zealand and for the country as a whole. These consumptions can be directly compared with the actual hourly observations of temperature, wind, and illumination as well as forecasts of these same parameters. In a study by the author,[4] half-hourly electricity consumption values were obtained for the Wellington City area (population 250,000), and for New Zealand (population 3 million) as a whole. However since all days are not equal from an energy consumption viewpoint (for example Sunday consumption is quite different from Wednesdays), it was considered to be more meaningful to compare the same day of each week, and Wednesday was taken to be the most 'normal' day of the week. The study considered the nine Wednesdays in the mid-winter months of July and August 1969.

Electricity consumption data for the Wellington area for each of the 24 half-hourly loads (e.g. 12.30 a.m. to 1.00 a.m., 1.00 a.m. to 1.30 a.m. etc.), were considered first and an analysis showed that the differences between the actual consumption and the average for specific half-hour periods exceeded 20% in some cases. For example on Wednesday 2 July 1969, the half-hourly differences varied from as much as 18% above average between 8.30 a.m. and 9 a.m., and 21% above average between 4.30 p.m. and 5 p.m.. In contrast, on Wednesday 13 August 1969, from 4.30 p.m. to 5 p.m., the consumption was 23% below average. A similar analysis of electricity consumption for New Zealand was made and a selection of the actual half-hourly differences from the average is shown in Table IX.1. Some of these differences are economically significant; for example, the demand from 1.30 p.m. to 2 p.m. on Wednesday 2 July 1969 was 240 megawatts (or 12%) above the average.

As previously noted, if the peak-demands could be forecast more accurately (using, for example, detailed temperature, wind, and illumination forecasts), the proportion of thermally generated electricity would be minimized. To assess the specific relationship between electricity consumption and temperatures, a comparison was made of the differences from the average of the half-hourly power consumption and the temperature differences from the average for the same period. This showed that in general the larger the negative departure of temperature, the larger the positive departure of electric power consumption and vice versa. For example, the analysis for Wellington City showed that electric power consumption varied up or down by about 16 megawatts over a half-hour period (compared with an average consumption of about 120 megawatts) whenever the temperature was either 2° to 4°C above average, or 2° to 4°C below average.

Table IX. 1. Electric power consumption on Wednesdays in New Zealand during July and August 1969. Differences (in megawatts) from the average Wednesday consumption during this period

Half-hour periods	Average	Successive Wednesdays								
		1	2	3	4	5	6	7	8	9
4.30 a.m. – 5 a.m.	930	+30	–20	0	+80	+20	–30	–40	0	–30
7.30 a.m. – 8 a.m.	2250	+180	+60	+40	+200	+80	–100	–80	–100	–260
9.30 a.m. –10 a.m.	2290	+180	+20	+50	+160	+90	–170	–210	–30	–50
1.30 p.m. – 2 p.m.	1980	+240	+60	0	+220	+40	–220	–230	–130	+50
5.30 p.m. – 6 p.m.	2520	+160	+50	+40	+110	+60	–100	–190	–90	–30
8.30 p.m. – 9 p.m.	2290	+160	+30	0	+100	+30	–110	–120	–60	–40

Source: After Maunder (1971c).

The relationship between hourly electricity consumption and hourly temperatures was also carried out for New Zealand in a similar way to that described for Wellington City, with the exception that the temperatures were weighted according to the consumer demand for electricity. The data showed that in July and August 1969 temperature departures in the 2° to 4°C (3° to 7°F) range were associated with variations in the electricity consumption of 150 to 200 megawatts, which represented about 10% of the average usage of electricity during these periods.

Weather conditions clearly have an important impact on electricity consumption. Specifically, the studies cited show that on 'cold' or 'warm' days in winter in New Zealand, the electric power consumption increases or decreases by a factor of 10 to 15%. The decision-making involved in the economical generation and supply of electricity is considerable, and there is strong evidence that the 'correct' use of temperature forecasts could prevent the unnecessary generation of more expensive electricity. In addition, prior knowledge of very cold periods could assist local supply authorities in estimating the peak demands.

3. Monitoring Electric Power Consumption: The Value of Weather Information to the New Zealand Decision-Maker

The New Zealand Meteorological Service has for several years been providing the central electricity generating authorities with forecasts of rainfall, temperatures, and wind, as well as detailed real-time analyses of population weighted temperatures. Special weather forecasts are issued every morning of the temperature, wind, and 'illumination' for noon, and 5 p.m. today, and 8 a.m., noon, and 5 p.m. tomorrow for eight urban areas, as well as rainfall forecasts for nine hydro-catchment areas. These forecasts are sent to generating system controllers and are used to assess both the demand for electricity during the next twenty-four hours, and the likely water flow into the hydro-catchment areas. A detailed analysis of the population-weighted temperature deviations from the average for 24 places in New Zealand at 8 a.m., noon and 5 p.m. for each day is also provided, the resulting temperature indexes being used by the electricity authorities for 'monitoring' and 'explaining' the hour-to-hour demand for electricity.

The value of this information to the electricity authorities in New Zealand is difficult to assess. However, since the weather is the dominant variable affecting domestic power consumption in New Zealand, it is estimated that in the absence of this weather information, the costs of electricity generation (in 1988) would be increased by at least US$2 million a year (equivalent to about 15% of the annual budget of the New Zealand Meteorological Service).

4. Summer Electric Power Consumption and Temperatures: A United States Example

Cooling degree days which are a measure of the 'cooling' necessary to cool buildings to a satisfactory level are computed in real-time by the National Weather Service in the United States[5] and provide useful information to decision-makers in the energy sector of that country, particularly during very hot conditions. Such information - particularly if it is commodity-weighted, by in this case the human population - may also be used to weather 'adjust' several indicators of United States economic activities. Specifically, the author in a 1982 study[6] suggested that the 'true' indicators of United States economic activity during the 1980 summer would have been lower than those published and, more importantly, would have given different week to week variations *if* the effect of abnormally high temperatures had been taken into account. Table IX.2 shows that the United States electric power production in a 14-week period in 1980 was 6.3% higher than in 1979. During the same period there was an increase in the population-weighted cooling degree-days from 844 to 1075, giving an *average* weekly increase of 16.5 cooling degree-days. Disregarding purely 'economic' factors, the data therefore indicated that for the 14-week period, the 27.4% increase in population-weighted cooling degree-days (from the same period in the previous year) was associated with a 6.3% increase in the total industrial, commercial, and domestic electricity consumption.

Of course there were also 'true' economic factors which caused electricity consumption to increase from 1979 to 1980. Indeed, projecting an analysis of United States electric power consumption for the period from mid-July to mid-August for the 12 years 1968-79 to the same period in 1980 indicated an 'expected' increase from 1979 to 1980 of about 3.8%. In comparison the actual average weekly power production in 1980 was 51,135 million kwh, and 48,618 million kwh in 1979, giving an actual increase of 5.2%. The difference was therefore 1.4% higher than expected.

Table IX.2 United States electric power production and cooling degree-days: summer 1980 and 1979 compared

Week ending	Electric power production (millions kWh)			Cooling degree-days (population-weighted)		
	1980	1979	1980/1979 (%)	1980	1979	1980/1979 (%)
June 28	46,894	44,256	106.0	83	46	180
July 5	45,838	42,232	108.3	77	48	160
July 12	49,165	46,206	106.4	90	78	115
July 19	52,635	47,691	110.4	110	75	147
July 26	49,943	48,066	103.9	84	87	97
Aug. 2	50,126	49,200	101.9	97	90	108
Aug. 9	51,834	49,516	104.7	105	85	124
Aug. 16	49,507	43,316	114.3	75	36	208
Aug. 23	47,981	45,679	105.0	73	64	114
Aug. 30	48,138	46,407	103.7	81	74	109
Sept. 6	46,844	44,980	104.1	75	67	112
Sept. 13	45,549	43,370	105.0	53	42	126
Sept. 20	44,434	41,239	107.8	39	25	156
Sept. 27	44,351	41,172	107.7	33	27	122
Total (14 weeks)	673,239	633,430	106.3	1,075	844	127.4
Average (14 weeks)	48,089	45,245		76.8	60.3	

Source: After Maunder (1982a).

As previously noted the population-weighted cooling degree-days for the same period in 1980 and 1979 showed an average weekly increase of 16.5 population-weighted cooling degree-days, or 27.4%. Some (or even all) of the increase in cooling degree-days was clearly directly associated with the assessed 'real' increase in the national electricity consumption of 1.4%. On a unit base therefore, it would appear that a 1% increase in the requirements for cooling (using as an index of cooling the national population-weighted cooling degree-days) is associated with a weather impact on the United States electricity consumption of 0.051%. Alternatively, a weekly increase of *one* population-weighted cooling degree-day could be said to be associated with a weather impact on the United States electricity consumption of 0.085%. While these percentages are small, their impact on local and regional electricity supply companies, in times of high summer temperatures when the demand for electricity for air-conditioning is at a premium, can be very significant.

A major component of the weekly indicators of economic activity published in the United States, such as those in *Business Week*, is electric power production. The relationship between these economic indicators and electricity consumption is discussed in Chapter X, but it is relevant to note here that since the electricity production component of the *Business Week Index* has a weighting of 17.3%, any increase or decrease in electric power production (from one week to the next) will have a significant impact on the national index of economic activity. It also follows that this will occur *irrespective of the reason* for the increase (or decrease) in electric power production. This is because economic indexes such as the *Business Week Index* are usually based on the premise that an increase in production (including electric power), or an increase in consumption (including electric power), automatically indicates an increase in 'economic activity'.

While such a premise is appropriate in most production and consumption activities it is suggested that it should not automatically apply to electric power production and consumption, since a proportion of the variation in electricity demand reflects weather conditions rather than economic 'strength'. In fact there would appear to be strong reasons why the normally accepted economic growth models in which increased electricity consumption (at least for domestic purposes) is automatically assumed to be related to economic strength should be questioned.

It is also presumably important to separate 'real' economic growth from that associated with a 'temporary' weather/climate variation. In fact, an increase in electricity consumption to cool or warm buildings (and the people inside them) could well be considered - at least from a national viewpoint - 'unproductive' except in the sense that it allows people to be more comfortable and hence be more productive. Naturally, any increase in electricity consumption means additional revenue for utility companies, and in those countries where electricity is produced by commercial companies rather than the government, the companies will rightly state that an increase in electricity consumption *from whatever cause* is to their economic advantage. However, from a national viewpoint the question must be asked as to whether such an increase is *always* an indicator of national economic prosperity? Indeed there would appear to be a good case for suggesting that increases in electric power production as a result of higher than average summer temperatures, or colder than average winter temperatures, should be considered 'negative' rather than 'positive' contributors to national economic activity.

5. Application of Weather Modification to the Electric Power Industry: Two Case Studies

In many countries there is an increasing demand for electricity as a result of the growth of air-conditioning in houses, offices, and factories. One interesting aspect of this was explored in a study[7] which considered the modification of temperatures by contrail clouds, and the possible effects of such modification on electric power consumption and production costs. In the analysis, electric power loads in the

United States mid-west and their associated temperatures were subjected to a modification scheme based upon the formation of contrail clouds. The costs of generating electric power under modified and unmodified temperature conditions were then used as a basis for providing an insight into the implications of temperature modification for the electric power industry.

An analysis given in the original paper, for the costs of producing power using a modified 85°F (29°C) series, shows that the *hourly* cost differences at that time varied from $2,027 to $50,425, with an average of $22,540. The results of the model suggested that the benefits of modification through contrail clouds could on very hot days be significant. Since this study was completed over 20 years ago, no known deliberate attempt has actually been made to create contrail clouds on any large scale, but the possibility exists if an intervention was considered to be economically and strategically desirable.

The most practical use of weather modification relates to hydro-electric power generation, and a test programme of the Pacific Gas and Electric Company operating (when the study was made) 67 hydro-electric plants with an installed capacity of 2,226 MW, is described in a study by Eberly.[8] He outlined some of the factors that need to be taken into account in assessing the economic value of weather modification and pointed out that if an increase of precipitation from cloud seeding could be depended upon, it would be possible to introduce efficiencies in hydro-electric systems, since water levels in reservoirs at the end of a dry season are usually maintained at a level which will provide a reasonable assurance that the lakes will be full by the end of the following wet season.

Investigations of the potential benefits from cloud seeding for several different watersheds assumed that 'increased precipitation' from seeding results in a 10% increase in runoff, and that the value of the increased runoff is in the fuel saved by not using thermal generation. Eberly indicated that if a 10% increase in precipitation from cloud seeding could be achieved, the benefit/cost ratios of various watersheds would vary from a low of 0.4:1 to as much as 14:1 depending on how wet or dry the season was.

B. MANUFACTURING / WEATHER RELATIONSHIPS

1. Perception of the Effects of Weather on Manufacturing

The results of an early report on weather and manufacturing suggested that in the United States,[9] manufacturing had the least economic benefit potential of any weather user classification studied. However, since an appreciable proportion of the national income of many countries is derived from manufacturing (e.g. Japan 36%, Thailand 18%, Australia 31%), more comprehensive studies of the specific impact of weather and climate, and weather forecasts on manufacturing would appear to be warranted.[10] In this connection, a study[11] of the perception of the effects of weather on manufacturing firms in Colorado is noteworthy. Two aspects of the problem were considered: the perceived and real effects of weather on various functional areas of manufacturing firms; and the present and potential utilization of weather information by such firms.

Five firms were selected for this study. First, a firm manufacturing precision mechanical components for the aerospace industry was examined. Production and quality control were the two operations most weather-affected, variations in temperature causing changes in dimensions and tolerances, and high humidity increasing the cost of product maintenance and storage. Second, a brewery was considered. Here the chief weather effects were related to the source of barley and rice. In addition, severe drought conditions severely restricted the plant's water supply, and freezing temperatures necessitated the use of insulated railroad cars for shipping. A brick and ceramic product manufacturing company was the third firm studied. In this plant, snow and cold presented a number of problems. For example, wet clay lowered the grinding efficiency, thus decreasing both output and quality, and a drop in temperature below -12°C necessitated a change from natural gas to propane, causing a substantial increase in cost in the operation of the kilns. The ski apparel industry was also considered, the firm being particularly weather sensitive to the marketing of their products. Finally a firm involved in consumer and industrial durables was studied. In this case a substantial proportion of the products sold were for use in the winter. Consequently a lack of snow, ice, and freezing conditions during late autumn, when a considerable amount of advertising occurred, seriously inhibited sales.

The study concluded that despite the initial minimal awareness of the effects of the weather on the five companies, the actual effects were found to be considerable, particularly in terms of costs. A programme to inform management that better decisions could be made if more appropriate use was made of weather and climate information and weather and climate forecasts would appear to be justified.

2. Impact of Severe Weather Conditions on Manufacturing Activities

Few in-depth studies have been made on the impacts of severe weather conditions on manufacturing, but a recent investigation[12] of British industry which evaluated the impact of a severe winter and a hot/dry summer on the 'performance' of industries is significant. Unfortunately, the unavailability (or non-use) of 'industry or commodity-weighted' weather information reduced the value of the results, but the survey revealed some interesting contrasts. For example, in 1962/63 severe winter conditions increased activities in the utility industries by 17% over the previous winter, whereas activities in the brick and cement industries decreased by 14%. The analysis also showed that a hot dry summer increased activities in the clothing, footwear, drinks, and tobacco industries, but decreased activities in the pottery, glass, and utility industries.

C. RETAIL TRADE / WEATHER RELATIONSHIPS

1. Some Previous Weather / Retail Trade Analyses

In a study[13] of weather and retail trade in 1950 it was suggested that the weather might affect the retail sales in four ways: it could be too 'uncomfortable' to go shopping; it could produce situations which would physically prevent people from going shopping, as in the case of snow drifts or floods; it may have a psychological effect on people in that they may change their shopping habits; and some kinds of merchandise may be more desirable (and hence saleable) during a period in which certain weather conditions occur.

Many other factors influence the demand for goods, but whether these are primarily economic, sociological, or psychological, weather conditions have a considerable overriding influence. A key factor is the 'urgency to shop', and this involves decisions related to both the actual and forecast weather, any alternatives to shopping, and the type of goods required. However, irrespective of the type of shops available, or the means of transportation to and from them, weather conditions and the seasonal climate do have a considerable influence on the type of goods sold, and the amount of profit made by a retailer.

In some cases the astute retail manager utilizes such knowledge in planning day-to-day sales, for by correctly assessing both the 24 to 36 hour and the 4 to 5 day weather forecasts, the day(s) on which there is likely to be the largest number of customers can be predicted. Such knowledge is also useful in determining the times during which maximum sales personnel should be available, and when it would be most suitable for holding unadvertised sales. Longer-range forecasts are also helpful, and any improvements in monthly and seasonal weather forecasts should enable those responsible for forward purchasing to become more efficient buyers, since such forecasts should be a guide to the kind and quantity of seasonable merchandise that is likely to be required by the customer in the near future.

The specific effects of weather on the daily sales of three departments of Younker Bros Inc., in Des Moines, Iowa, were the subject of an early study by Steele.[14] The results indicated that in the period studied (the seven weeks before Easter from 1940 to 1948) 88% of the variability in the sales of the store were accounted for by the weather. In another study, Zeisel[15] examined the importance of weather on beer consumption in Rhode Island and found that for every 1.0°F (0.5°C) temperature change, the consumption of beer changed 1.1% above or below the expected level.

Several studies were compiled by Linden for the U.S. National Industrial Conference Board, and in one paper[16] the effects of 'adverse' weather on customer traffic in department stores are discussed. In the analysis, weekly data for New York City for 1957 and 1958 were used. Linden noted that after appropriate adjustments for non-comparable weeks, sales and weather moved in the 'same direction' four times more frequently than they moved in the 'opposite direction'.

That is high sales and 'good' shopping weather were four times more likely to occur than high sales and 'bad' shopping weather.

The effects of weather on the total retail trade of a country are also of considerable importance, and can in fact pose difficult problems for economists trying to explain variations in certain key economic indicators. One major problem is 'explaining' the variations which take place from 'seasonally adjusted' data, and Petty[17] in commenting on the movement in the Federal Reserve Bank of Chicago index of department stores sales stated:

> We had to make some judgement as to how much of the poor showing to blame on (1) severe weather or (2) the consumer's persistent reluctance to buy. The relatively sharp drop-off in sales in areas hardest hit by heavy snows and extremely low temperatures suggested that abnormal weather was again exerting its influence on consumer buying. As someone has remarked: 'Most seasonal adjustments for last February "took out" only 2 inches of the 10 inch snowfall we had'.

2. A Japanese Example

Although the weather clearly affects retail trade, few in-depth studies have been made; indeed the need to cite the above studies from the 1950's,1960's, and the 1970's reflect the sparse nature of published material linking weather and retail trade. One exception is the study by the author on weather and United States retail sales[18] which is discussed in the next section; another is the informative survey by Gabe[19] on *Weather Information - Valuable Economic Tool in an Era of Low Growth* published in the August 1985 issue of the Mitsubishi Corporation's *Tokyo Newsletter*. The survey provides a valuable insight into how weather information is used in day-to-day decision-making as the following extract shows:

> One leading brewery in Japan keeps data collected over the past 30 years showing the relationship between growth in beer consumption and the weather index. ...The 'brewery's beer weather index' is compiled by recording daily maximum temperatures in 15 areas around the country where its branches are located. ... When the maximum temperature registers one degree centigrade above normal on a fine day in July or August, beer sales that day increase by 2,470,000 large-size bottles. This is equivalent to an increase of about 8% over the average daily consumption.

Gabe points out, however, that it is very difficult for this brewery to incorporate weather information into its planning process, and quoting a member of the brewery's planning department, he states:

> Beer consumption is closely related to weather. And yet, it is not easy to plan beer production based on weather forecasts. Beer is not something that can be

stocked for a long time; like fish, it must be supplied fresh. The important thing for us is to establish business operations that are not affected by the weather.

The demand for air-conditioners is similarly affected by the weather. The *Tokyo Newsletter* says that air-conditioning data in Japan revealed that weather became a major determining factor around 1977-78. Specifically, the intense summer heat in 1978 was responsible for the sale of 450,000 more units than the 2,700,000 anticipated, whereas in 1982, a cooler summer than usual, only 1,900,000 units were sold compared with expected sales of 2,400,000.

3. A United States Example

The formulation and application of weighted indices on a weekly basis specifically designed for United States retail trade is discussed in this special example.[20] The factors of 'large area' (i.e. the United States) and 'short time period' (i.e. a week) were deliberately chosen for two reasons: first, it is considered that nation-wide economic activities are important to various 'high-level' decision-makers, and second, only weather over a short period has any real practical meaning to the millions of 'low-level' decision-makers who each day go shopping in the United States.

Various weekly economic indices are available for the United States including those published (until 1979) by the Department of Commerce's *Weekly Retail Sales*. These data formed the basis of the investigation. However these data were available *only* on a nation-wide basis and restricted to 13 major stores types. Accordingly, no 'official' weekly information were (or are) available relating to weather- sensitive commodities such as ice cream, soft drinks, women's winter and summer clothes, refrigerators, air-conditioners, automobile tyres, beer, iced tea, paint, umbrellas, etc.

In order to use the available nation-wide retail sales, it is necessary to compute a weekly weather index for the United States. An immediate problem is that people thinking of going shopping, or actually shopping, react to the weather as a whole rather than to the individual weather elements of rain, snow, temperature, wind, humidity, or sunshine. The combination of weather elements into a single useful retail-related weather index is difficult, and a further consideration is that the only (then available) suitable published weather data on a weekly basis was that in the *Weekly Weather and Crop Bulletin* (a joint publication of the U.S. Department of Agriculture and National Weather Service). Weather indices in the form of weighted weekly precipitation and temperature indices were therefore computed, bearing in mind that an index compiled from an appropriate combination of wind, sunshine, humidity, temperature, and precipitation duration would have been more appropriate.

The nation-wide weighted weather indices were evaluated from weekly temperature and precipitation differences from the average for 147 places across the United States. These places had a total population of 120,000,000. A weighting was computed for each of the 147 places based on the 'buying power index' published in the *Marketing Magazine*. An abridged example of the calculation of the United States weighted temperature and precipitation differences from the average for the week ending 11 February 1968 is given in Table IX.3. The study covered three years, nation-wide precipitation and temperature indices for the United States being computed for each of the 154 weeks from April 1966 to March 1969.

Table IX.3 Abridged example of the calculation of the temperature* and precipitation** departures from the normal for the United States for the week ending 11 February 1968, weighted according to the 'buying power index' of 147 localities

No.	Station	Buying power index (US = 100)	Departure from normal temp. (deg. F)	precip. (in.)	Weighted differences temp. (deg. F)	precip. (in.)
1	Birmingham, Alabama	0.32	−9	−1.3	−2.8	−0.42
15	Denver, Colorado	0.62	+4	−0.2	+2.4	−0.12
19	Washington, DC	1.67	−2	−0.5	−3.3	−0.84
22	Miami, Florida	0.63	−8	−0.2	−5.0	−0.13
32	Chicago, Illinois	4.22	−2	−0.3	−8.4	−1.23
50	New Orleans, Louisiana	0.50	−11	−1.0	−5.5	−0.50
65	Kansas City, Missouri	0.75	−3	−0.3	−2.3	+0.23
72	Las Vegas, Nevada	0.14	+3	+0.1	+0.4	+0.01
80	New York, NY	9.27	−4	−0.8	−37.1	−7.42
122	Dallas, Texas	0.79	−1	−0.6	−0.8	−0.47
132	Salt Lake City, Utah	0.26	+4	−0.3	+1.0	+0.08
147	Cheyenne, Wyoming	0.03	+7	−0.1	+0.2	+0.00
	Total (147 stations)	63.12	–	–	−152.8	−35.35

Source: After Maunder (1973).
Notes:
* Total weighted temperature departure = −152.8
  Average departure = −152.8/63.12   = −2.4 deg. F (−1.3 deg. C)
** Total weighted precipitation departure = −35.35
  Average departure = −35.35/63.12   = −0.56 in. (−14.2 mm)

The association between retail trade sales and the weighted precipitation and temperature indices for the United States such as those shown in Table IX.4 were assessed using a regression equation. Technically this allowed an assessment of the relationship between variations in retail trade sales from a time trend over specific 11-week periods, and the variations in the weighted nation-wide precipitation and temperature indices. The results of the analysis indicate several nation-wide features: (1) drier conditions than average appear to be associated with above average retail sales in late winter and early autumn; (2) wetter conditions than average appear to be associated with above average retail sales in late spring and early summer; (3) colder conditions than average appear to be associated with above average retail sales in the autumn; and (4) warmer conditions than average appear to be generally associated with above average retail sales in spring and early summer.

There are, of course, many other aspects of weather and retail trade, including the permanency of lost sales. That is, how permanent is the loss of business experienced on bad shopping weather days, or 'is the sale that is lost today gained tomorrow?' Some evidence[21] suggests that a specific item not bought is not always purchased when better shopping weather comes along. More importantly, 'bad' weather means fewer trips to the shop, a factor which is usually associated with

Table IX.4 Selected weekly retail trade sales* in the United States for the period January to March 1968 and associated precipitation and temperature indices weighted according to the buying power index of 147 localities

Period	Retail business (millions of dollars)					Weighted indices**	
Week ending Saturday	Total retail trade	Apparel group	Furniture and appliances	Lumber building hardware	Drug stores	Precip. (in.)	Temp. (deg. F)
Jan. 13	5,344	284	250	227	205	+0.10	-8.5
Jan. 20	5,562	296	281	255	215	+0.10	+1.9
Jan. 27	5,581	273	276	264	208	-0.28	+1.8
Feb. 3	5,706	277	295	275	203	+0.28	+6.4
Feb. 10	5,720	292	287	286	210	-0.56	-2.4
Feb. 17	5,772	264	278	297	227	-0.43	-5.8
Feb. 24	5,778	284	302	287	200	-0.48	-6.3
Mar. 2	6,049	275	295	315	207	-0.25	-2.8
Mar. 9	5,967	306	291	316	214	-0.16	-1.7
Mar. 16	5,967	306	271	330	199	+0.74	-1.6
Mar. 23	6,121	327	279	342	206	+0.17	+0.8
Mar. 30	6,548	376	304	376	196	-0.51	+7.5

Source: After Maunder (1973), data from Weekly Retail Sales Report - U.S. Department of Commerce/ Bureau of the Census.
Notes:
* Weekly sales estimates are based on data from 2,500 firms, covering approximately 48,000 retail stores in the United States.
** Weighted weather indices are for each week ending midnight on the Sunday of the week indicated. The indices shown are departures from the normal for the specific week. Data are not adjusted for seasonal or holiday variations.

lower total buying, since a single exposure of the buying impulse is usually less rewarding - from a retailer's point of view - than two exposures.

Analyses of retail sales which consider only weather as a contributing factor omit, of course, much more important factors that determine the desire and ability of consumers to buy, namely the ability to pay, anticipated future income, price expectations, and present ownership of goods. Nevertheless, as has been shown the weather can explain a significant part of the variability in the sales of many items.

D. ROAD CONSTRUCTION/ WEATHER RELATIONSHIPS

1. The Setting

Road construction firms need information on the number of work days possible, both for bidding on road construction contracts and scheduling machinery and man-power, while contract-letting officials need such information to anticipate correctly probable completion dates, and to schedule funds for payment to contracting firms as work progresses. Answers to such questions are not usually available from traditional climatological analyses, but in 1970 a comprehensive study[22] developed a 'soil' moisture index and applied it to engineering data to estimate conditions suitable for work in the road construction industry of Missouri.

The study estimated the hours of construction time available in various calendar periods, the frequency of the different kinds of 'work-weeks', and the relationship between the types of work-weeks and the manpower and machinery requirements for specific road construction activities. It also formulated an index designed to estimate progress on construction projects on a regional basis. The analyses showed that after translating weather information into operational values, the relationships needed to convert operational data into costs could be developed.

2. The Effects of Weather on Road Construction: A Simulation Model

The effects of the weather on highway construction were estimated through their influence on working conditions. Data from two construction projects were linked with a 'soil' moisture index to assess the conditions under which construction activities could proceed. This relationship generated a series of working conditions based on the available weather data, and was then used as an aid to the planning and scheduling of highway construction projects.

Operational data for a road construction job in Missouri were obtained from the Missouri State Highway Commission and two private contracting companies. Using these data, each day during a sample four- year period was classified by the resident engineer into one of three categories: a full workday, a no-work day, or a partial workday. Saturdays, Sundays, and holidays were excluded unless work happened to be done on those days.

Development of an appropriate 'soil' moisture index required two types of information: daily soil moistures for a period comparable to the period over which the construction data were available; and information on the ability to operate heavy equipment in adverse soil moisture conditions (that is the 'trafficability' of equipment).

Typical computations of the soil moisture index are given in Table IX.5. This shows that on 2 June 1966, the soil moisture index at the end of that day was 1.20 in. (30 mm) and that the maximum soil moisture index loss on the following day would be 0.11 in. (2.8mm). Similarly, the soil moisture loss on 5 June was 0.07 in (1.8 mm), giving a soil moisture of 1.06 in (26.9 mm) at the end of that day. A similar day-to-day analysis of the soil moisture index for all days in a 50-year period from January 1918 to December 1967 was made.

The soil moistures were then linked with the construction data to produce a series of simulated working conditions. Four 'workday' categories were defined: a holiday, Saturday, or Sunday, when no work was to be done (symbolized by 0); a normal workday without any 'weather' restrictions (symbolized by 1); a no-work day due in the main to adverse weather/soil moisture conditions (symbolized by 2); and a partial or restricted workday (symbolized by 3).

Table IX.5 Examples of soil moisture index computations

1966 date	Soil moisture index previous day (in.)	Maximum possible loss (in.)	Precipitation on day n (in.)	Is precipitation ≥ maximum loss?	Actual soil moisture loss (in.)	Soil moisture index on day n (in.)
May 24	1.56	0.16	1.61	yes	0.16	1.80
May 25	1.80	0.16	0.00	no	0.13	1.67
May 26	1.67	0.16	0.00	no	0.11	1.56
May 27	1.56	0.16	0.00	no	0.09	1.47
May 28	1.47	0.16	0.00	no	0.07	1.40
May 29	1.40	0.16	0.00	no	0.06	1.34
May 30	1.34	0.16	0.00	no	0.05	1.29
May 31	1.29	0.16	0.00	no	0.04	1.25
June 1	1.25	0.16	0.00	no	0.03	1.22
June 2	1.22	0.16	0.00	no	0.02	1.20
June 3	1.20	0.11	0.17	yes	0.11	1.26
June 4	1.26	0.16	0.00	no	0.13	1.13
June 5	1.13	0.11	0.00	no	0.07	1.06
June 6	1.06	0.11	0.00	no	0.06	1.00
June 7	1.00	0.11	0.18	yes	0.11	1.07

Source: After Maunder, Johnson, and McQuigg (1971a).

To apply the workday classifications to the daily weather record, an analysis was made comparing the rainfall and the computed soil moisture index for several hundred days in the 1965-68 period, with the *actual* workday classification for two road construction jobs. It showed that it was possible to estimate correctly the workday classification by considering only the daily precipitation and the computed soil moisture index. Using this procedure the soil moisture series was translated into a working day classification (Fig. IX.1).

One direct application of the simulation model is the calculation of an 'index of workability', essentially an index of whether work can or cannot be done. The 'workability index' employed was calculated by assuming the time available for work in a full workday to be 8 hours, a partial workday 4 hours, and a no-work day 0 hours. These values are arbitrary but they are a reasonable approximation of what actually occurs. The workability index was computed for Jefferson City, Missouri for all days in the 50-year period 1918-67, the various values for the main construction months April to October being given in Table IX.6. These data show that on average 70-80% of the possible time could have been worked, the monthly average varying from 69% in April and May to 82% in July. The highest monthly

Figure IX.1 Graphical relationship between soil moisture index and work classification at Jefferson City, Missouri, summer 1966

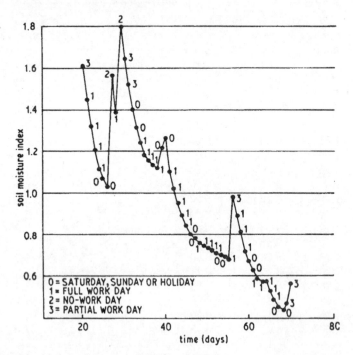

values for the 50-year period were all above 90%, and in seven months they were 97% or higher. By contrast, in April 1922 only 38% of the possible work could have been done.

Table IX.6 Workability index for Jefferson City, Missouri, 1918-67

Month	Highest (year)	Mean (s.d.)*	Lowest (year)	Range
April	0.92 (1956) (1959)	0.69 (0.12)	0.38 (1922)	0.54
May	0.94 (1934)	0.69 (0.12)	0.42 (1935) (1938)	0.52
June	1.00 (1936)	0.71 (0.15)	0.45 (1935)	0.55
July	0.97 (1940)	0.82 (0.09)	0.60 (1961)	0.37
August	0.97 (1960)	0.80 (0.09)	0.61 (1952)	0.36
September	0.98 (1939) (1940) (1950)	0.79 (0.12)	0.48 (1926)	0.50
October	0.98 (1964)	0.79 (0.11)	0.39 (1941)	0.59

Source: After Maunder, Johnson, and McQuigg (1971a).
Note: * Standard deviation.

The estimated monthly workability index for two years is shown graphically in Fig. IX.2, 1936 being a dry year for road construction, 1924 a wet year. Similar graphical analyses were obtained for all years, and a graph showing the variation in the work index in May for the 41-year period 1918-58 is shown in Fig. IX.3. Both graphs show the variation in the work that could have been done using all days including weekends, and the difference between a wet and dry year, and a wet and dry month, is very evident.

3. The Effects of Weather on Road Construction: Applications of the Simulation Model[23]

Questions for which the road construction industry needs answers include: how many hours of work are available during the normal construction season, and what kind of variation can be expected; if the construction project is started at a particular time, and a certain number of hours or days is required for its completion, can the job be completed by a pre-specified date; if several jobs are

Figure IX.2 Comparison of work index values for the wet year 1924 and the dry year 1936, April-October

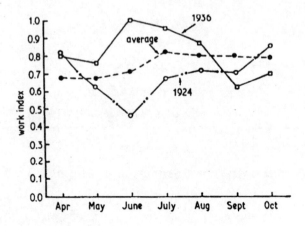

Source: After Maunder, Johnson, and McQuigg (1971a).

Figure IX.3 Work index value variations in May at Jefferson City, Missouri, May-July 1918-58, if weekends are included as workdays

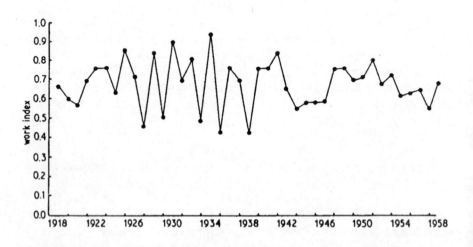

Source: After Maunder, Johnston, and McQuigg (1971a).

initiated at different starting dates, and all require an equal amount of work, when will each one be completed; if a job is started on a particular date, and the succeeding period is 'unusually wet', what is the probability of being able to complete the job in the prescribed time; and at what point in a 'wet' road building season does it become virtually impossible to complete a specific construction job on a schedule?

To provide answers to these questions, the daily series of working conditions were converted into an index expressed in hours of worktime; as previously noted, this was done by weighting 'full workdays' by 8 hours, 'partial workdays' by 4 hours, and 'no-work days' by 0 hours. The number of hours available for work during specific periods can be summarized in various ways. For example, a useful method of assessing the influence of working conditions on worktime is to assume a fixed amount of worktime (say 1,000 hours) and a variable starting time. In any construction period there will be a 'run of days' that will be the shortest period of time during which the required work can be accomplished. There will also be a 'run of days' that will be the most unfavourable for road building, and thus represent the longest period of time that would be required to complete the desired amount of work. Sample values of the type just described are shown in Table IX.7 for every fifth year in the period studied. These data indicate that a road building project in

Table IX.7 Sample values of the range of cumulative calendar days required to complete 1,000 hours of worktime (weekends included). These values are based on rainfall data for Jefferson City, Missouri.

Year	Minimum time		Maximum time	
	Starting date	Calendar days to completion	Starting date	Calendar days to completion
1920	July 5	157	March 26	176
1925	July 9	158	June 2	171
1930	February 27	141	July 14	155
1935	June 30	149	February 25	194
1940	June 30	140	February 27	168
1945	July 11	155	February 26	204
1950	June 13	142	March 11	166
1955	July 25	140	February 27	169
1960	June 14	138	February 20	164
1965	July 11	157	April 11	181
Sample average	June 20	148	March 30	174

Source: After Maunder, Johnson, and McQuigg (1971b).

Missouri requiring 1,000 hours of work has the most chance of completion if it is started in late June or early July. In contrast, the same job started in late February or early March is likely to experience the greatest delay.

It is evident that the road construction industry could benefit from specific research on the effects of weather on their activities. But more carefully designed records describing specific progress on construction projects are needed, as well as on-site measurement of weather events. This would allow the industry better to identify and measure quantitatively the impact that the actual weather has on the industry, as well as the value of weather information.

E. TRANSPORT / WEATHER RELATIONSHIPS

1. Productivity and Transport

Given sufficient lead time, the value of accurate weather-based forecasts of a nation's production is wide-ranging, involving many activities 'beyond both the farm and factory gate'. The key factor in optimizing the value of these forecasts is the need for a realization that weather and climate variations *can* be taken into account. This is now taking place in the United States, among other countries; indeed, in 1977 the U.S. Secretary of Agriculture in looking at the agricultural sector of the U.S. economy issued a ruling[24] that '... never again do I want to see an economic report come out of the (U.S.) Department of Agriculture that assumes average weather'.

This 'average weather' and 'average climate' misconception is regrettably still a feature of many production industries - and by inference the associated transport industries. Indeed, many decision-makers within the transport industry appear to be quite content to go from month to month and year to year with an almost complete disregard to the effects of weather variations on their operations.

2. Transport Costs Beyond the Farm Gate: A New Zealand Example

New Zealand's success as an exporter of meat, wool, and dairy products is highly dependent upon her ability to produce high quality products at relatively low cost. Indeed Flemming,[25] writing in a publication of the New Zealand Ministry of Agriculture and Fisheries, stated that with the blessing of a favourable climate and efficient farming, New Zealand can produce meat at the farm gate for a fraction of the energy input used in Europe or the United States. However he sounded a note of warning when he commented:

By the time that meat has been processed at a freezing works the energy content has more than doubled and a further doubling occurs between works and market. ... It is astonishing to realise that the direct energy used in producing, processing and transporting a lamb carcass (from New Zealand) to the United Kingdom is equivalent to half its weight in oil.

The meteorological input into minimizing these energy costs, and in particular those associated with the transportation sector, is a significant factor in maintaining New Zealand's traditional competitiveness in meat, wool, and dairy products.

The fluctuations in New Zealand's pastoral production from season to season also has significant linkages to transportation since most of this production is exported. In this regard Troughton[26] made some pertinent comments:

A major problem is to reconcile the highly seasonally-dependent production with processing and market demand. This approach to management introduces the need to consider the whole agricultural chain from the farm to the market place as one integral system.

The month-to-month variations in the meat, wool, and dairy industries are also significant. In the case of lamb slaughterings in New Zealand, Taylor[27] noted that one of the main requirements for an efficient production, processing, and marketing chain is the accurate forward estimation of the volume of produce which the chain is required to handle. He further commented, 'As a country so dependent on exports and hence transport can we afford to see our major export earner function at anything less than peak efficiency through the whole "pasture to market" chain?' The implications of not providing appropriate meteorological advice relating to the whole 'pasture to market' chain are clear.[28]

F. CONSTRUCTION / WEATHER RELATIONSHIPS

1. Impact of Weather on Construction Planning

The effects of weather and climate on the construction industry, especially in those countries in which the winters are severe, include variations in the rate of employment of workers, increased construction costs, and the degree of risk and uncertainty in the planning of construction projects. The weather and the climate have economic impacts at three levels - on the contractor, the construction worker, and the customer. Since no two jobs are exactly alike, it is not possible to evaluate the economic impact of weather and climate on the construction industry in any neat, closed form; however, its effect on a specific job can be estimated by the use of a simulation model. Such a model[29] was presented at a conference in St. Louis in 1971.

The hypothetical job was a two-storey office building comprising 41 activities of different durations and sensitivities to various weather elements. In a specific case, a job was started on 1 April with an assumed level of reliability of weather predictions of 90%. The total estimated job duration with no allowance for weather-caused delays was 101 days, but adverse weather conditions increased considerably both the job duration and labour costs. Specifically, 21 out of the 321 decisions required in the job were incorrect because of the adverse weather, and this resulted in 20 extra days' work being required. In addition, there were 'indirect' costs associated with the delay in starting future jobs.

Many more questions were created than answered during the development of this simulation model, and the authors commented that only a more realistic model will determine what specific information the construction manager needs in order to make his daily decisions, and what it is worth to collect and disseminate this information to the construction industry.

2. The Economic Impact of Weather on the Construction Industry of the United States

The United States Weather Bureau sponsored a study[30] in the early 1960's to determine the nature and magnitude of losses due to the weather in the construction industry of the United States. The potential capability of present and future weather forecast accuracy, as well as the use of meteorological and climatological services to reduce these losses was also examined. The resulting analysis indicated that the total annual loss due to weather varied from 3% to 10% of the value of construction.

Among the conclusions reached by the study was that if specific weather and climate information *had been made available* to the construction industry *and* correctly used, a saving of 15% of the weather and climate caused losses was possible. In addition, the savings achievable, if the weather forecasts for the 0-24 hour time periods were 100% accurate, were estimated to be worth 30% more than those obtainable with the then forecast accuracy.

3. Applications of Probability Models in Using Weather Forecasts to Plan Construction Activities

The inherent variability of weather contributes substantially to the uncertainties in the costs of building. This point was emphasized by Johnson and McQuigg in a study[31] which showed that the use of historical weather data and information typically contained in construction 'logs' can be employed to develop probability models. Such models indicate the likelihood that alternative activities can be undertaken, and are based upon the observed and forecast weather conditions.

The study also pointed out that construction managers could use such models to make decisions relating to who to call to work, and which construction activities should be active. The cost of making decisions in this manner - which would include paying for the appropriate weather information - could be more than offset by savings in construction costs.

G. NATIONAL INCOME AND PRODUCT ACCOUNTS:
A NEW ZEALAND EXAMPLE

A more thorough and orderly process of assessing national climate sensitivity can be obtained through an analysis of national income and product accounts. But, although the trained eye may well be able to differentiate between sectors of an economy which are more weather and climate sensitive than others, it is not easy. Consider for example how the climate sensitivity of New Zealand can be assessed through examining the various components of the economy as they appear in the *New Zealand Official Year Book*. (The method used may provide an example as to how a similar 'econoclimatic view' could be applied to the equivalent statistical surveys of other countries.)

The approach used involved an overview of the economy and then a more detailed look at key elements. A few examples will indicate the typical step-by-step process required to make an initial assessment of national climate sensitivity.

In terms of agriculture one of the key factors in assessing 'absolute' rather than 'relative' climate sensitivity is to consider agricultural production in terms of its dollar value. In New Zealand, wool accounts for about 19% of the gross value of agricultural production, compared with 3% for vegetables. Thus although some aspects of the vegetable sector (such as transporting vegetables) are much more weather sensitive than many aspects of the wool industry (such as the effects of severe frost on the quality of wool), in *monetary* terms the value of the weather and climate to the total national agricultural scene would have a much greater contribution from the wool sector than the vegetable sector.

Similarly, since 62% of New Zealand's agricultural income is from pastoral products (i.e. wool, dairy products, and meat) it is evident that the key climate factors in New Zealand will be very much related to the state of the nation's pastures. Naturally in other countries, this would be quite different, especially where field crops or horticultural type crops are the principal agricultural earners.

Further, consider the components of the gross domestic product (GDP). In New Zealand, 11% comes from agriculture and 6% from food manufacturing. Other important weather and climate-related production groups include transport and storage (6%), construction (5%), energy (3%), and wood products and forestry (3%). These components comprise one-third of New Zealand's gross domestic product, and since they are much more weather and climate sensitive than other sectors, they must be considered key weather and climate components of national income accounts.

The weather and climate sensitive sectors of other nations will generally be quite different from those of New Zealand. However, an analysis of similar national 'Year Books' may provide an initial assessment of what are the significant weather and climate sensitive sectors. What has been described relates to a 'feel for the situation' rather than the results of an in-depth analysis however, and illustrate only one aspect of the 'weather-economic' mix. Few in-depth studies have been made of the total mix - particularly on a national scale - and the analysis of the New Zealand economy while providing an initial answer to the problem is far from complete, and much work remains to be done.

Chapter X

FORECASTING PRODUCTION

A. THE ASSESSMENT ROLE

Irrespective of the weather observations made, and the associated economic analyses done, the question to be answered is 'what does it all mean?' To answer this question, appropriate assessments must be made using all available information, and to do this skilled assessors with vision and knowledge are required.

Several such assessors are active in a number of countries and a noteworthy aspect of their work is the important and not usually understood concept that 'although last week's or last month's weather is history the assessment and measurement of its economic impact is certainly not history'. As previously noted, this occurs because economic 'indicators' (such as the consumer price index) take several weeks (or months in some cases) to compile, whereas meteorological data collection systems provide worldwide information on the state of the atmospheric environment within minutes of the 'events' taking place. Thus, providing appropriate research has been undertaken, real-time weather information can provide first estimates of that part of the economy that is weather sensitive. Accordingly, a forecast of economic data such as production or prices - *which at this moment do not exist* (or more correctly have not been collated or forwarded to any data collection system) - can be made using real-time weather information.

In addition, since weather and climate are generally considered to be 'outside of the economic system' their direct effects are easier to measure than the effects of other economic variables. Further, since weather and climate-induced trends are often part of an economic trend, it is evident that if they are not correctly explained they may well mask the economists', agriculturalists', and politicians' understanding of a system or process, thereby preventing correct decisions being made. The inclusion of real-time weather and climate information in any economic assessment should therefore provide a fuller understanding of many economic 'events'.

B. WEATHER / ECONOMIC MODELS

1. An Overview

Weather/economic models fall into two broad categories with regard to the degree to which the models consider 'physical' processes. Models that do *not* include all the salient physical-biotic mechanisms (such as those used in most weather/ production studies) are described as being 'empirical'. That is they employ a convenient but logical framework in order to correlate predictor and predicted variables, but they are not specifically based on any *real* economic-physical-biotic mechanisms, because those mechanisms are either not known, or not known in sufficient detail to be used. In contrast, 'explanatory' models endeavour to specify accurately and explicitly the real relationships between economic/climate components and processes.

2. Explanatory Models

The non-availability of relevant data is the main reason why explanatory models have been much less developed by applied climatologists, despite the fact that their utility is potentially much greater than empirical models. The purpose of an explanatory climate model is to specify cause-and-effect or stimulus-and-response relationships between climate components and processes. Correctly developed explanatory models can therefore be used to predict not only climate events (and by inference the associated economic impacts), but also variables in past or future environments for which actual observations are not available. Explanatory models are used to considerable success in studies of the consequences of the predicted increase in 'greenhouse' gases on atmospheric circulation, temperature, and sea levels through the statistical 'manipulation' of numerical atmospheric circulation models.

3. Empirical Models

Empirical models by contrast employ a convenient but logical 'framework' so as to correlate predictor and predicted variables. Some of the current empirical models indicate a variety of useful relationships between a range of economic variables and weather data for differing time scales. In addition, they assist in developing more useful space/time orientated agrometeorological models which can be used in real-time weather-based agricultural production forecasting.

Many subtle relations between climate conditions and human activity await identification and description. Scientists often view the empirical type of research and its results as only 'first' estimates of reality, for clearly the models do not 'explain' processes. However the statistical relationships resulting from such models very often condense complex cause-and-effect computations into simpler computations that reduce the cost of evaluating the model. In addition, in many cases - particularly in the area of agricultural production weather relationships for 'large' regions - empirical models are currently the only models which can be used operationally.

C. WEATHER AND CLIMATE BASED FORECASTS OF AGRICULTURAL PRODUCTION

1. An Overview

Relationships between agricultural production and climate are usually assessed through the development of agroclimatological models. Some of these models attempt to mirror the physical processes that actually cause variations in agricultural production (such as soil water availability), but in many cases a statistical relationship is established such as linking rainfall and temperature variations with variations in wheat or rice production. These statistical models do not 'explain' environmental processes, but they do condense complex cause-and-effect linkages into useful associations.

In general, climate and agricultural production models assess the significance of a weather/climate variation on either the productivity of 'human-altered agricultural environments' such as the Iowa corn field or the Korean rice paddy, or the productivity of the more 'natural' environments of the Australian merino sheep or the Wyoming range cattle. The economic, political, and strategic importance of being able to predict the significance of larger-scale weather and climate fluctuations on agricultural production also needs to be considered, particularly as this information contributes to national and global security.

2. Dairy Production Forecasts: A New Zealand Example[1]

New Zealand, the world's largest exporter of dairy products, markets and sells its products in more than 100 countries. In this regard, the quality, and price of the products are very important, but of equal importance is the guarantee of a regular supply of specialized dairy products to an increasingly food-conscious world. This regular supply is largely dependent on the quantity and quality of production from the dairy farm; forecasts of production are therefore of paramount importance to decision-makers at all levels in the New Zealand dairy industry.

Milkfat processed by dairy factories in New Zealand varies from year to year (e.g. 275, 251, and 274 million kg in the successive seasons 1976/77, 1977/78, and 1978/79) and also from month to month (e.g. less than 1 million kg in June, to 35-40 millon kg in most Octobers, Novembers, and Decembers). The most important variations occur in the January to May period and in particular in the autumn month of March, but of greater significance is the difference between the actual production and the 'optimal' or 'ideal' production.

Various methods can be employed to estimate these differences. One solution is the use of appropriate commodity-weighted weather and climate information (specifically in this case weighted by the population of dairy cows) in a statistical regression type analysis, in which differences from season to season (as shown in Table X.1 for the March to March differences) are used.

Table X.1 Milkfat processed by dairy factories in New Zealand in March, and the percentage changes from March to March

Year	Production (million kg)	Value* ($ million)	Production (% of previous year)
1976	24.1	108	114
1977	22.8	103	96
1978	17.4	78	76
1979	23.7	107	136
1980	28,0	126	118
1981	23.0	104	82
1982	24.9	112	108
1983	22.6	102	91
1984	30.8	139	136
1985	31.3	141	101

Source: Updated from Maunder (1984a), compiled from information supplied by the New Zealand Dairy Board.
Note:
* Based on a value of $4.50/kg which is the value in May 1984 if milkfat is converted into cheese and exported (1985 value about $5/kg).

Table X.2 Data for 'dairying region'* analysis - New Zealand

Season	Weighted days of soil water deficit**			Dairy production***	
	Dec.	Jan.	Feb.	Feb.	Mar.
1966–67	−0.6	0.0	−0.2	107	105
1967–68	0.0	+1.7	+12.8	92	77
1968–69	0.0	−1.1	−12.1	113	136
1969–70	+0.5	+4.7	+18.7	76	50
1970–71	+1.9	−2.0	−15.5	108	160
1971–72	−2.3	−0.1	+7.8	107	104
1972–73	+1.3	+2.1	+8.5	96	74
1973–74	−0.5	+9.9	−6.2	75	100
1974–75	0.0	−11.9	−7.7	140	130
1975–76	−0.5	−2.6	−3.9	119	115

Source: After Maunder (1977a).
Notes:
* North Auckland, South Auckland, Bay of Plenty and Taranaki Dairy Board areas. Approximately 85% of the total New Zealand milkfat production comes from these areas.
** Differences in days from previous year for same month.
*** Percentage of previous year for same month.

Soil water deficit data are a convenient way of 'converting' the vast amount of weather and climate data into information which can be meaningful to studies relating weather and climate to dairy production. Typical 'difference' data, for both days of soil water deficit and dairy production, are shown in Table X.2 for the main dairying area of New Zealand for selected months in the period 1966/67 to 1975/76.

The first operational use of the predicting model was in the 1972/73 season; it has now been successively updated each season to include more 'seasonal history' in the model, so that predictions for the 1987/88 season were based on data for the 36 seasons 1950/51 to 1986/87. The current weather-based dairy production models used in New Zealand are based on the differences in the (dairy cow) weighted number of days of soil water deficit, rainfalls, and temperatures. Differences in these indices (from season to season on a monthly basis) are used in various statistical equations to predict dairy production as a percentage change from the corresponding month(s) in the previous season. In order to quantify these relationships various statistical regression equations are used.[2]

In practice it has been found that one particular equation is the most useful. This equation allows a prediction of production differences in say March (early autumn), using only the weather data differences for the previous three months (ie., December, January, February). Further, if (in this case) the weather conditions in February can be forecast, then the actual prediction for March could be done as early as mid-January. Other equations can be used giving even more lead time. Indeed, a reasonably successful forecast can be made of April (mid-autumn) production, using only January (mid-summer) weather information, which in some cases may be of greater value than a more 'accurate' forecast given at the beginning of April.

The importance of providing production forecasts with sufficient lead-time to enable appropriate decisions to be made in regard to transportation and marketing[3] cannot be over-emphasized. A summary of the progressive forecasts made during the 1977/78 season is given in Table X.3 and this shows that the dramatic 'drop' in production in February, March, and April of 1977/78 (compared with the same months of 1976/77) was well anticipated. For example, with an 11 weeks' lead-time, the March production was predicted to be 69%, compared with the actual production of 76%. The significance of the relatively long lead-time predictions provided by this type of analysis is well illustrated by comparing the cumulative production (actual and forecast) during the season. Table X.4 gives such information for the 1977/78 season. These and similar weather-based forecasts of production are of considerable potential if not actual benefit to the decision-makers involved with the New Zealand dairy industry.

The value of rain or the cost of continuing dry conditions to the New Zealand dairy industry can also be assessed from this type of dairy production model, since the model uses the differences in the weighted days of soil water deficit from March to March, April to April etc. and each 'day of deficit' has a specific value in terms of production and value. An example is given in Table X.5 which shows

Table X.3 Predicted milkfat production in New Zealand during the 1977/78 season, with lead-time (LT) (in weeks) before actual production was known

Month of prediction†	Oct. %	Oct. LT	Nov. %	Nov. LT	Dec. %	Dec. LT	Jan. %	Jan. LT	Feb. %	Feb. LT	Mar. %	Mar. LT	Apr. %	Apr. LT
Jul.	101*	(15)*												
Aug.	103	(11)	103	(15)										
Sep.	101	(8)	105	(11)	105	(15)								
Oct.	106	(3)	104	(8)	105	(11)	102	(15)						
Nov.			102	(3)	103	(8)	99	(11)	104	(15)				
Dec.					104	(3)	102	(8)	107	(15)	108	(15)		
Jan.							101	(3)	76	(8)	69	(11)	71	(15)
Feb.									74	(3)	63	(8)	68	(11)
Mar.											82	(3)	57	(8)
Apr.													55	(3)
Actual production	98		102		99		95		81		76		46	

Source: After Maunder (1978).
Notes:
† End of month shown.
* i.e. Prediction was made at end of July 1977 for the New Zealand October 1977 milkfat production to be 101% of the actual production in October 1976. The actual production of 98% was not known until late November, giving a lead-time of 15 weeks.

the 'value' of rain in March 1977 in terms of the following months milkfat production. Based on the actual days of soil water deficit in January and February and the first 14 days of March, the model showed among other things that if no rain fell during the rest of March, then the April 1977 production would be 60% of the April 1976 production.

The model further indicated that rain sufficient to 'saturate' the soil would have effectively increased prodution by about 4% *for each day* that the soil retained moisture sufficient for grass growth. For example, if 'saturation rains' had occurred on the 25 March, the predicted April production would have increased from 60% to 82%, whereas 'saturation rains' 10 days earlier on the 15 March would have nearly doubled the predicted April production from 60% to 119%. The monetary value of this rain (see Table X.5) indicates that an extra 10 days of grass growth on the 'national dairy farm' from 20 to 30 March, would have been worth about 5.67 million kg (6.51-0.84) of milkfat or about NZ$14 million (in 1977 dollars and prices).

The use of successive models (for example, using January only, January plus February, January plus February plus March to predict April production) also provides valuable information. A summary of typical progressive dairy production

Table X.4 Cumulative predicted and actual milkfat production in New Zealand during 1977/78 in million kg

Month of prediction†	*Known production 77/78*	*Predicted production 77/78*	*Known + predicted production*	*Actual production 77/78*	*Actual production 76/77*	$\frac{A}{C}$ *(%)*	$\frac{B}{C}$ *(%)*
			A	B	C		
Oct.	*Oct.* –	*Oct.–Jan.* 159.6	159.6	*October to January* 151.0	153.0	104	99
Nov.	*Oct.* 39.5	*Nov.–Feb.* 142.0	181.5	*October to February* 172.6	179.7	101	96
Dec.	*Oct.–Nov.* 80.1	*Dec.–Mar.* 128.3	208.4	*October to March* 190.0	202.5	103	94
Jan.*	*Oct.–Dec.* 117.7	*Jan.–Apr.* 82.2	199.9	*October to April* 197.2	218.1	92	90
Feb.	*Oct.–Jan.* 151.0	*Feb.–Apr.* 44.8	195.8	*October to April* 197.2	218.1	90	90
Mar.	*Oct.–Feb.* 172.6	*Mar.–Apr.* 27.5	200.3	*October to April* 197.2	218.1	92	90
Apr.	*Oct.–Mar.* 190.0	*Apr.* 8.6	198.6	*October to April* 197.2	218.1	91	90

Source: After Maunder (1978).
Notes:
† End of month shown.
* e.g. At the end of January, the known production (October to December) was 117.7 million kg, the predicted production (January to April) was 82.2 million kg, giving a 'known plus predicted' production for the October to April period of 199.9 million kg. This is 92% of the production for the same period of the previous season, and compares with 90% which was the actual percentage not known until late May 1978.

forecasts made for April 1978 is given in Table X.6 which shows the various assumptions that were made. Most predictions (even the early ones) show that the 'dramatic' drop in production in April 1978 compared with April 1977 was well anticipated.

Of course, weather-based forecasts such as those described are not always accurate. The reasons for the errors are partly an incomplete understanding of the weather/biological processes, partly the unavailability of historical milk production data for shorter periods than the month, partly the influence of other factors (such as the 'drying off' of dairy cows and the rate of fertilizer application) which are only partially related to weather and climate factors, and a whole host of economic, social, and political factors which influence not only the farmer, but also it seems on some occasions even the cows.

Table X.5 Value of rain in March 1977 to New Zealand dairy industry in April 1977

Date of rain to saturate soil	Predicted April production* %	Effective change from that predicted on March 14		
		%	m.kg	$m.**
March 15	119	+59	+9.85	+24.6
March 16	113	+53	+8.85	+22.1
March 17	109	+49	+8.18	+20.5
March 20	99	+39	+6.51	+16.3
March 25	82	+22	+3.67	+9.2
March 30	65	+5	+0.84	+2.1

Source: After Maunder (1977b).
Notes:
* Percentage of 1976 production of 16.7 m.kg.
** Based on 1977 value of $2.50/kg which is value if milkfat is converted into cheese and exported (1985 value about $5/kg).

As of 14 March 1977 the April 1977 production was predicted to be 60 per cent of the April 1976 production, but would have increased if any rain occurred during March. Specifically, the April 1977 production could have increased to 119 per cent of the 1976 production, if rain equivalent to 17 days (31−14) of soil moisture (approximately 50 mm) occurred in the area. The value of this rain decreases with time from data of prediction (14 March) as shown in the table.

Table X.6 New Zealand milkfat production predictions for April 1978

Date of prediction	Model*	Assumption	Predicted production differences**	
			%	$ million
Jan. 9	J	No further rain in Jan.	−30	−12
Jan. 31	J	None	−35	−14
Feb. 1	J + F	Normal Feb.	−19	−7
Feb. 1	J + F	As Feb. 1977	−27	−11
Feb. 1	J + F	No rain in Feb./Temp. +1.5°C	−54	−21
Feb. 1	J + F	No days of soil water deficit/Temp. −1.5°C	−14	−55
Feb. 15	J + F	No further rain in Feb.	−52	−20
Feb. 15	J + F	No further days of soil water deficit in Feb.	−31	−12
Feb. 28	J + F	None	−45	−18
Mar. 1	J + F + M	Normal Mar.	—32	−13
Mar. 13	J + F + M	No further rain in Mar.	−69	−27
Mar. 31	J + F + M	None	−62	−24

Source: After Maunder (1986b).
Notes:
* J = January (i.e. model used January weather data only); J + F = January + February (i.e. model used January and February weather data).
** April 1978 compared with April 1977.

Actual difference was −54%, 'worth' $21 million at 1978 prices.

3. Wool Production Forecasts: A New Zealand Example[4]

One of the requirements for an efficient production, processing, and marketing system in the New Zealand wool industry is the accurate forward estimation of the quantity and quality of wool. There is a need therefore to establish a methodology for accurately estimating - preferably one season in advance - the volume of wool likely to become available.

A specific 'customer' for such forecasts is the Economic Service of the New Zealand Meat and Wool Boards which has to predict annually, with in-seasonal adjustments, national wool production.[5] These estimates are necessary for making farm income, production, and export volume assessments for the New Zealand Meat Producers Board and the New Zealand Wool Board, and such forecasts are also of value to other organizations in New Zealand such as the Institute of Economic Research and the Reserve Bank. Until recently estimates of New Zealand's wool production were based largely on 'subjective expert opinion', but recently econoclimatic models using differences in the weather from one season to the next have been used for predicting the national wool production in the subsequent season. Variations in various aspects of the wool industry in New Zealand are quite considerable, Fig. X.1 showing the season to season fluctuations in New Zealand's sheep population, wool production, and wool production per sheep from 1965/66 to 1983/84.

In New Zealand, studies[6] have shown that it is the weather and climate factors that affect the liveweight of sheep that are of prime importance, and in the econo-climatic models of wool production developed for New Zealand, the soil water deficit and the temperature deviations from average proved to be useful indicators of overall weather conditions. For example, when the soil water deficit index is high there is less moisture available for grass growth; therefore during such periods sheep could be expected to be relatively poorly fed. However, there is a consider-able time lag between the lack of soil moisture available for pasture growth and the associated linkage to the growth of wool. This lag is of the order of several months; indeed, an autumn drought has not only an effect on the feed available for sheep at that time, but also a latent effect on both lambing percentages and wool production in the following season since the sheep will 'go into the winter' in a poor condition.

The weather-based wool production models use the differences in both the sheep population weighted number of days of soil water deficit and the mean tem-peratures, these differences then being used to predict various aspects of wool production as a percentage change from the previous season. Typical days of soil water deficit data used in these models are given in Table X.7, and typical wool 'production' variations are shown in Fig. X.1. In order to quantify these relation-ships various statistical regression models are used involving various combina-tions of monthly weighted weather data for each of the 12 months in the season *before* production. For example, data for the months June 1986 to May 1987 would be used for predicting production in the 1987/88 season.

Figure X.1 Fluctuations in New Zealand sheep population and wool production:1965/66 - 1983/84 (data give percentage of the previous season)

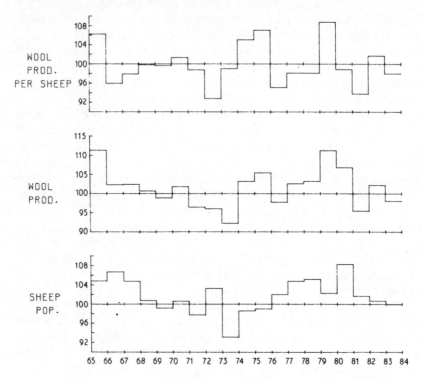

Source: Compiled from data supplied by the New Zealand Meat and Wool Boards' Economic Service.

The current weather/wool model includes data for the individual months June to May of the season prior to being forecast and associated combinations of these months, with wool production per sheep in the following season. An example of this model (using data for December 1977, and January, February, and April 1978 to forecast wool production per sheep for 1978/79 as a percentage of the production in the previous season) indicates that using weather information up to April 1978, the predicted wool production per sheep for the 1978/79 season would have been 96.5% of the 1977/78 season. Specifically, the model predicted that the late summer drought in 1978 (January and February 1978 each had 9 *more* days of soil water deficit than January and February 1977) would reduce the 1978/79 wool production per sheep by about 3.5% from that produced in 1977/78. In this model, weather information is used up to April 1978; the lead time is therefore 13 months prior to the end of the 1978/79 season.

The *actual* wool production per sheep in 1978/79 was 98.1% of the 1977/78 production; therefore the weather-based forecast made in May 1978 of 96.5% was reasonably accurate. The same model has been used to predict the wool production

Table X.7 Weighted* soil water deficit indices - New Zealand 'sheep farm'

Season	Jan.	Feb.	Mar.
1975/74	−6	−3	−3
1976/75	−3	−3	+11
1977/76	−5	+1	−2
1978/77	+10	+9	+5
1979/78	0	−6	−15
1980/79	−9	−5	−2
1981/80	+8	+7	+7
1982/81	+1	+1	−2
1983/82	−2	+1	+7
1984/83	−1	−9	−11
1985/84	+1	+5	+6

Source: Updated from Maunder (1980a).
Note:
* Weighted by the distribution of sheep. The index measures differences between seasons in days (e.g. January 1978–January 1977 = +10 days).

per sheep for the seasons following 1978/79, and Table X.8 gives details of the weather-based predictions of wool production and wool production per sheep for the 1980/81 to 1984/85 seasons, as well as the actual production data.

Forward estimation of the supply of agricultural production is a challenge that many organizations have. More specifically, people involved with the production, transport, processing, and marketing of wool need to know the likely availability, quality, and flow of wool internationally. Historically this information has been provided after a subjective assessment of the various factors which influence the quantity and quality of wool. The accuracy of these forecasts is largely dependent on the skill and experience of the individuals or groups making them, but if they are not biologically and/or weather based they can at times give misleading information, with adverse consequences to a variety of decision-makers both on and off the farm, including those outside of the country of production.

The weather/wool production models described have been designed to provide weather-based estimates on the likely level of wool production in New Zealand. They 'explain' a significant proportion of the annual variation in wool production, but they are by no means perfect. Used in association with the other more traditional methods they provide more soundly based forecasts, but an important additional factor to note in this regard is that there are few if any weather-based models of the *quality* of wool available, and the ideal wool/weather prediction model should be concerned with both the *quantity* and the *quality* since both factors influence the final price.

The quantity and quality of wool produced in the main wool producing countries - notably Australia, USSR, New Zealand, China, Argentina, South

Table X.8 New Zealand wool production: predicted and actual*

Date of prediction	Season	Predicted total wool production	Predicted wool/sheep	Actual total wool production	Actual wool/sheep
May 1980	1980/81	105	97	107	99
May 1981	1981/82	96	98	95	93
May 1982	1982/83	104	99	102	102
May 1983	1983/84	100	98	98	98
May 1984	1984/85	102	104	99	103

Source: *Actual* data compiled from information supplied by the New Zealand Meat and Wool Boards' Economic Service, and the Department of Statistics.
Note:
* All data expressed as a percentage of the previous season.
Data on the actual wool production per sheep and the total actual wool production is known at the end of each season, usually with a delay of about one month. Thus as of May 1984, when a weather-based forecast of the wool production was made for the 1984/85 season, the actual production was known for the 1982/83 season, and provisionally available for 1983/84, the final production for the 1983/84 season not being available until August 1984. The final production for the 1984/85 season was not known until August 1985, 15 months *after* the date of prediction.

Africa, and Uruguay - is the result of a combination of many factors. In this industry, which is vital to many of these countries, it is important to develop more efficient forecasting and monitoring techniques if costly errors of judgement are to be prevented. What is required is greater efficiency in the whole industry (specifically in this case wool, but obviously also applicable to many other commodities). This applies especially in the area 'beyond the farm gate', if the comparative agricultural advantages which a number of nations have 'enjoyed' in the past are to continue.

D. WEATHER AND CLIMATE BASED FORECASTS OF BUSINESS ACTIVITY

1. Business Activities: Sensitivity Analysis

The sensitivity to weather and climate of the business sectors of most countries is often more difficult to assess than those of the strictly agricultural sectors. However, several sectors - notably energy, transport, and trade - are potentially as weather and climate sensitive as are the more obvious agricultural sectors. Weather and climate fluctuations can produce significant economic, social, and political consequences, and since some aspects of productivity and consumption are ultimately associated with the availability, use, and interpretation of information on the past, present, and future weather and climate, there are strong reasons for certain weather/business linkages to be given greater recognition.

One specific aspect of the climate-economic mix is the importance of having 'leading economic indicators' available in near real-time, and both presidents and prime ministers and their advisors have found that such information provides an essential background for their day-to-day decision-making. This is particularly so in Europe and the United States but in most countries of the world there is a growing appreciation of the need for accurate and up-to-date economic data on national productivity and related matters.

As previously discussed, the traditional response of the climatologist to 'weather-interpreting' economic data has been 'luke-warm'; however, today several national meteorological services not only observe, collect, and process meteorological data in real-time but *also* analyse it in real-time on a commodity by commodity basis. This means that key commodity-related weather and climate indices can be computed and made available in real-time on a national and international basis. Clearly, this provides new opportunities for the traditional meteorologist and climatologist.

A specific example of the value of such economic-related indices[7] is their use in assessing the influence of weather on a leading United States indicator of economic activity, namely 'housing starts'. The results indicated that in critical spring months, such as April, the effect of weather fluctuations on housing starts can have significant impacts on the whole housing industry.

The availability of 'climate productivity indices' for key business commodities is now a reality. Further, the availability of such information through instant information and retrieval systems offers a tremendous challenge not only to meteorologists and climatologists in terms of the marketing potential of their services, but also to the decision-makers who have the potential to use such information. One specific aspect of these new products is their use in 'weather-adjusting' national economic indicators.

2. National Economic Indicators: A United States Example[8]

Many business activities rely to a considerable extent on the compilation of national weekly, monthly, and seasonal business indicators. This is particularly so in the United States and one advantage of looking at that country is that its sheer size, and the importance of the private sector, means that there is a real need for such indicators. Several nation indicators are published in the United States for periods as short as a week. The availability of such short-period indices not only allows relevant weather-impact studies to be made for the United States, but also permits the transfer of such experience to other countries. Indeed many countries find that collations of national economic indicators for as short a period as a week are essential if 'close monitoring' and/or 'fine-tuning' of an economy is to take place either by governments, or the private business sectors. In these situations, there is a need to provide clear guidance to economists and decision-makers on what specific impacts the weather and climate has had, or will have on economic activities.

In the United States, *Business Week* computes and publishes each week their *Business Week Index*, and a similar index of national economic activity is published by the *U.S. News and World Report*. Both indices are computed by weighting 'seasonally adjusted' economic activities such as raw steel production, rail freight traffic, and electric power production. The *Business Week Index* includes twelve components, the important electric power production component comprising a high 17.3% of the total weighting.

A comparison of the index for the summers of 1979 and 1980 show average values of 150 in 1979 and 138 in 1980 (the base 1967 = 100). Part of the decrease of 12 points from 1979 to 1980 reflected the downturn in the United States economy that occurred at that time, but of greater significance are the variations from week to week, and in particular (in this case study) the trend within each summer. Although the downward trend was greater in 1980 than in 1979, there is strong evidence that the index would have decreased *even more* rapidly if electricity production in the 1980 summer had not been at record high levels. These high levels of electricity production were primarily the result of the extremely hot conditions in the 1980 summer, and it is clear that they kept the index 'artificially' high. This is because the index assumes (not necessarily logically) that an increase in electric power consumption *from any cause* - including the use of air-conditioners in a heat wave - is directly related to an increase in national economic activity.

During the 1980 summer, the *Business Week Index* varied from 139 to 136 (1967 = 100) with quite significant week to week differences. These variations together with the weekly electric power production expressed as a percentage of the comparable week(s) in 1979 are shown in Fig. X.2. Electricity production is a significant factor in the *Index* and a question to be asked is 'Would the "downturn" in United States economic activity (in the 1980 summer) have been more pronounced if the heat wave (which was the main cause of the significant increase in electricity consumption during this period) had *not* occurred?' More importantly, since the severity of the heat wave was well known meteorologically in real-time, and indeed forecast several weeks in advance, could that knowledge have been used to monitor more correctly and even predict United States economic activity?

It is considered that this is an excellent example of how actual and forecast weather information could be used to monitor and forecast an economic index, well *before* the official trend in the index is known. For example, actual temperature data - which are available in real-time - could have been used to adjust the actual electric power production data *before* they were incorporated in any economic index. In addition, a correctly used forecast of temperatures (assuming the forecast was a 'useful' indication of likely conditions) could be of considerable value in forecasting that part of the economic index which is weather sensitive, well before the actual index is known.

Figure X.2 The *Business Week Index* and U.S. electric power production (1980 as a percentage of 1979), June-September 1980

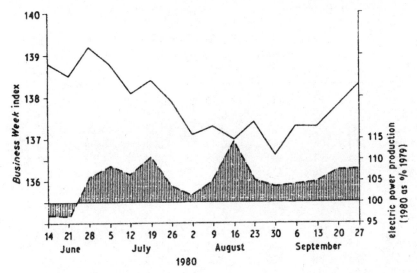

Source: After Maunder (1982a).

3. *Adjusting the* Business Week Index

As discussed it is considered that the 'business indexes' of United States economic activity should be adjusted to take into account those weather factors which 'artificially' increase the important electric power production component of such indexes. Such an adjustment for the *Business Week Index* was briefly discussed in Chapter IX, where it was shown that for each 1% increase in the requirements for cooling (above that expected on the basis of average temperatures), the *Index* needed to be reduced by a factor of 0.009%. It was also shown that for each population-weighted cooling degree-day (above the average for the specific week), the *Index* needed to be reduced by a factor of 0.015%. The specific application of these adjustment factors is now examined.

As an example, consider the impact of the percentage increase in the population-weighted cooling degree-days for the United States for the week ending 13 September 1980. These totalled 36% more than the average for that week, and the published *Index* was 137.3. Using the above adjustment factor, it is considered that this index should therefore be decreased by a factor of 0.32%(i.e. the 0.009% 'adjustment factor' x 36). The weather-adjusted *Business Week Index* for that week would therefore have been 136.9 or 0.4 index points lower than that published. Similar adjustments were made for the other summer weeks of 1980, the published and adjusted indices being shown in Table X.9 and Fig. X.3.

Table X.9 *Business Week Index:* percentage weather adjustments

Week* ending 1980	Population-weighted cooling degree-days				Business Week Index		
	1980**	Normal	% Normal	Weather adjustment (%)	Published	Weather- adjustment	Weather- adjusted index
June 28	83	63	132	−0.29	139.2	−0.4	138.8
July 5	77	69	112	−0.11	138.8	−0.2	138.6
July 12	90	73	123	−0.21	138.1	−0.3	137.8
July 19	110	76	145	−0.41	138.4	−0.6	137.8
July 26	84	77	109	−0.08	137.9	−0.1	137.8
Aug. 2	97	76	128	−0.25	137.1	−0.3	136.8
Aug. 9	105	73	144	−0.40	137.3	−0.5	136.8
Aug. 16	75	72	104	−0.04	137.0	−0.1	136.9
Aug. 23	73	64	114	−0.13	137.4	−0.2	137.2
Aug. 30	81	55	147	−0.42	136.6	−0.6	136.0
Sept. 6	75	47	160	−0.54	137.3	−0.7	136.6
Sept. 13	53	39	136	−0.32	137.3	−0.4	136.9
Sept. 20	39	32	122	−0.20	137.8	−0.3	137.5
Sept. 27	33	25	132	−0.29	138.3	−0.4	137.9

Source: After Maunder (1982a).
Notes:
  * Week ending Saturday.
** Week ending Sunday.

The overall effect of the weather adjustments may be described as the 'weather impact'. For example, Table X.9 shows that in the consecutive weeks ending 9 August and 16 August 1980, the difference in the 'published index' was -0.3 (137.0 compared with 137.3), whereas the difference in the 'weather-adjusted index' was +0.1 (136.9 compared with 136.8). Thus the impact of the percentage weather adjustments on the published *Business Week Index* is to change it from a *decline* of 0.3 to an *increase* of 0.1.

The major reason for this difference was the very hot conditions in the week ending 9 August 1980, in which the population-weighted cooling degree-days were 44% above the average for that week, compared with the near average conditions for the week ending 16 August 1980 in which the cooling degree-days were only 4% above the average. Indeed a near record electric power consumption of 51,834 million kWh occurred in the United States for the week ending 9 August, whereas in the following week electric power consumption was closer to the average at 49,507 million kWh.

The conclusion, therefore, is that if the published *Business Week Index* of 137.3 and 137.0 for the two successive weeks had been 'weather adjusted', it would have

Figure X.3 The *Business Week Index*, as published and weather adjusted (top), and (bottom) actual and average U.S. population-weighted cooling degree-days, June -September 1980

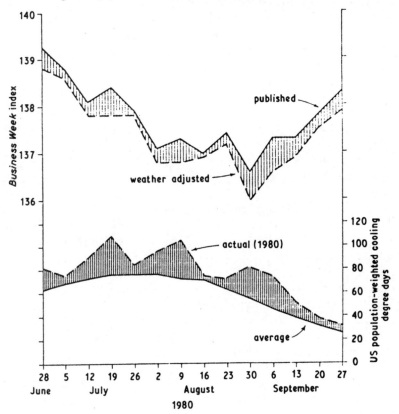

Source: After Maunder (1982a).

shown values of 136.8 and 136.9, indicating a slight weather-adjusted increase and *not* the 'published' decrease in overall business activity. The *Index* may therefore have given a 'false' impression of the direction of the United States economy in the week ending 16 August 1980.

The impact of the actual change (rather than the percentage change) in the national population-weighted cooling degree-days may also be assessed. Considering the week ending 13 September 1980, the actual national population-weighted cooling degree-days of 53 were 14 above the average for that week, and the published *Business Week Index* was 137.3. If this index has been decreased by the appropriate factor (i.e. the 0.015% 'adjustment factor' x 14) or 0.21%, then the weather-adjusted *Index* for that week would have been 137.0 or 0.3 points lower than that published. This may be compared with a reduction of 0.4 points if the percentage change, rather than the actual change, was considered.

It is evident that weather adjustments of weekly indicators of economic activity are significant, particularly in their ability to 'fine tune' the trend more correctly in national business activities. For example, although the published comment on United States economic activity between the weeks ending 6 September and 13 September 1980 was 'unchanged', a more accurate picture reflecting the dramatic change in the weather between those two weeks may well have been 'strengthened'. It is considered that 'weather-based fine tuning' is very important in correctly assessing economic trends.

4. The Way Ahead

A further refinement of weather-adjusting economic data is that it is possible to use real-time commodity-weighted weather information to provide a forecast of the tendency of national economic indicators. Indeed, since weather information is available in real-time, whereas most national economic indicators have a considerable time publication delay, weather-based forecasts of economic activity can be made available well in advance of the availability of 'official' production/ consumption information.

Indeed, it cannot be over-emphasized that while last week's or last month's weather is history, the *measurement* of its economic impact using conventional economic techniques takes a considerable time. But a forecast of economic data, which at the time the forecast is made have yet to be compiled, collected, or analysed, can be made using the unique availability of real-time weather data. Thus, where and when weather is a limiting factor, its economic impact can be assessed immediately.

Attempts to 'weather-adjust' national economic indicators are difficult and controversial; nevertheless such attempts not only provide economists, decision-makers, and publishers of economic indicators with realistic adjustment factors, but they also provide additional 'guidance' for the economic forecaster. It is also believed that appropriate analyses of the weather-economic mix can contribute positively and usefully in assisting decision-makers to understand better some of the world's important problems associated with the supply and utilization of food, fibre, and energy.

Traditional guidance to the economic forecaster comes from a variety of sources, and in a special issue of *Climatic Change* (Vol.11, Nos 1/2,1987) guest editors K.C. Land and S.H. Schneider state that social and natural scientists are often called upon to produce or participate in the production of forecasts. They therefore assembled in this issue of *Climatic Change* various papers that (1) describe the organizational and political context of applied forecasting, (2) review the state-of-the-art for many forecasting models and methods, and (3) discuss issues of predictability, forecasts errors, model construction, linkages, and verification. They further comment that among the issues common to many of the papers are (1) the question of whether theories that apply to small scales of time or space can be inflated to larger scales with success, (2) the advantages and disadvantages of complex models as opposed to simpler methods such as informed extrapolation, and (3) the fact that forecasts invariably involve some elements of human judgement and occur within a political context.

An associated aspect of forecasting, and in particular forecasting productivity related to the business environment, is the difficult question posed in a Review Essay in *Climatic Change* (Vol. 12, No.1, 1988) by Thomas Levenson who asks 'How much do we know, and how much should we tell?' He comments that these are two questions that afflict any science journalist trying to translate technically complicated information into stories for the public. He adds that '... science journalists function as gatekeepers regulating, by conscious design or as an unintended consequence of their own ignorance or lack of sophistication, what it is about any scientific endeavour that the public gets to learn.'

The 'gatekeepers', as Levenson describes science journalists (and presumably business journalists as well), have I believe an additional responsibility in providing information - including forecasts - to the right people at the right time. What is 'right' in any context is often difficult to assess, but the 1988 summer drought in the United States provided an ideal opportunity for journalists, editors, *and scientists*, as shown by the following examples from the American press:

USA Today 20 June 1988
> The farm economy will need two years to recover from widespread drought unless rain comes soon. ... A projection of the drought's impact by economist Stan Hargrove of DRI/McGraw Hill Inc., finds farmers will face lower income ... more debt ... and stagnant land values ... .

Wall Street Journal 21 June 1988
> Corn and soybean future prices opened at the upper limit of their daily trading ranges and stayed there for the entire session as drought fears dominated the Chicago commodity markets. ... Meanwhile, rumours swept the commodity industry that spiraling grain prices were choking off export sales of grain. Analysts said a scheduled sale to China of 500,000 metric tonnes of soft wheat for October delivery fell through yesterday.

While most aspects of forecasting production are related to the immediate future, there is also concern over production in the twenty-first century if, as many climatologists predict, there is a significant global warming with associated precipitation and storm pattern changes. Nevertheless, it is important to appreciate that irrespective of any climate *change*, climatic *variations* will remain. All countries will be subject to these variations and it would be very unwise for any country to place all their 'climate eggs' in the 2030 or 2050 basket, and not be equally or even more concerned with what will go into and out of the basket during the next 30 to 50 years.[9] As always we have to live within our climatic income.

Chapter IX

THE FUTURE

A. THE WEATHER ADMINISTRATOR OF THE FUTURE[1]

1. The Ministry of Atmospheric Resources

It is a sunny 15 March 1994. Welcome to another episode in the award-winning series 'A Day in the Life of the Decision-Maker', an intimate, in-depth look at what is going on in our government corporations. It is brought to you by New Zealand Independent Television's Channel 15. It is live, unscripted, and unrehearsed.

This time we're going to drop in on Dr Brown, Administrator of the Ministry of Atmospheric Resources (MAR). It is one of our newest and most important agencies. MAR was set up in 1990. It now has a budget of over $160 million and a technical staff of 1,000. We know it best as 'the agency that brought the atmosphere in from the cold', and which gave it status along with other resources such as water, agricultural land, forests, and fisheries. It has brought an enlightened approach to resources management in New Zealand. It is seen by similar agencies overseas as bold and imaginative. Little wonder that MAR now regards itself as the 'the agency that looks upon the nation from above'.

The Ministry of Atmospheric Resources consists of four main branches. These are responsible for Atmospheric Intelligence, Atmospheric Marketing, Atmospheric Management, and International Affairs respectively. Each is in the hands of a Director who oversees its overall management and who reports to a Corporate Affairs Group headed by the Administrator of MAR.

The Ministry is a very practical body. You can depend upon it for an accurate forecast of the weather, not merely for the next week but also for the next three months. It helps you plan your ski trips and your Pacific Island vacation. It gives financiers up-to-date, weather-based economic indices for key areas - both within and outside of New Zealand. It determines the effects of waste disposal into the atmosphere and issues permits to factories and to car owners for effluent discharge. It also collects the fees. Besides this, MAR is a referee. Its Clouds Jurisdiction Division sorts out disputes between individuals who have attempted to modify the weather and those who claim compensation from those who have done so.

The International Affairs Division provides New Zealand's input into the New International Meteorological Organization (NIMO), which was established in 1990 following the break-up of the UN-based World Meteorological Organization. It also deals with atmospheric resource disputes with other countries, and between the State Governments of North and South Zealand.[2]

We take you now to the office of Dr Brown, located on the 7th floor of the Ad Astra Building. This magnificent edifice on the hills overlooking Kapiti Island[3] was opened only in 1989, as part of the government's Decentralization of Head Offices Programme. Buildings, you see, now move to the staff rather than vice versa.

The Administrator is talking with Dr Jones, his Chief Executive Officer:

Dr Brown. Good morning, Dr Jones. What have we got on the agenda today?

Dr Jones. It looks like a very busy day. Lots of interesting items: some of them quite urgent.

Dr Brown. All right then - what's the first[4] item?''

2. Seasonal Forecasts and Climate Impacts

Dr Jones. Due to be released today is the forecast for winter 1994, and spring 1994. The forecast for winter indicates severe cooling in eastern areas of New Zealand, and a significant reduction in the rainfall in the major hydro-electric catchment areas. These forecasts will be disseminated through the usual channels, including the new environmental television and videotex service.

You will recall that seasonal forecasts for rainfall, temperature, and soil moisture have been made operationally since autumn 1988. Since then their track record has been very good - so much so that there is an embargo on the release of today's seasonal forecast until 10 p.m.. This will allow the Special Intelligence Unit of the Prime Minister's Department to make appropriate recommendations to the Prime Minister over what, if any, part of the forecast should be restricted.

Dr Brown. Does it look as though we have got a class A seasonal forecast for issue tonight, or is it likely to be unrestricted?

Dr Jones. Well, the nation will certainly face problems if the forecast is correct - and the Special Long Range Forecasting Unit is very confident. What it indicates is that there is a 90% probability of a second spring season in succession with very cold temperatures. It is expected that these cold conditions (following on the cold and dry autumn which we are now experiencing) will reduce the 1994/95 and 1995/96 wool production by as much as 25% with a significant impact on the fertilizer and shipping industries.

In addition, a substantial increase in unemployment is anticipated (mainly as a result of the extremely harsh growing conditions which are now entering their second season). As a result the Minister of Employment will have to be advised. Besides this, the Environment Planning Council will be called into emergency session.

The National Energy Authority in its meeting tomorrow will indicate to the 50 local energy authorities that their requests to the North and South Zealand State Governments for energy must be increased by at least 30%. Because of the financial difficulties that this will place on domestic users, the Ministers of Finance and Domestic Affairs will be advised that provision for supplementary benefits to the local authorities must be provided for in the 1994 budget.

Dr Brown. It looks pretty serious then, especially if we also take into account the atmospheric dust problem.

Dr Jones. It sure does. In fact, this dust problem is partly the reason for the cold conditions being forecast. It looks as though we could be reaching a crisis point over this. I have prepared a brief for the Minister reminding him that, while atmospheric carbon dioxide has continued to increase during the last decade, it is now growing extremely rapidly. This has resulted from the serious and wide-spread forest fires in Africa and South America in the early 1990's. However, and this is the significant factor, the volcanic eruptions[5] of Mount Etna, el Chichon and Krakatoa, all within three weeks of each other in 1993, has meant that the expected significant warming through carbon dioxide accumulation has been replaced (at least on a temporary basis) by an extremely cold spell in the Northern Hemisphere.

It seems that this cold spell is now spreading into parts of the Pacific. This has been shown in the cold summer, and the already cool autumn. The forecast cold winter and spring emphasize it even more.

Dr Brown. Presumably, we will have to call a media conference about this?

Dr Jones. Well, in addition to the items I have already mentioned in regard to wool production, shipping, employment, and energy, I have been in touch with the Minister of Health. It seems we will need to alert the Long Range Forecasting Unit, and the Climate Monitoring Unit to provide a confidential review to the Minister of the likelihood of the cold conditions continuing next summer and even into next autumn. The health problems associated with continued cooling could be quite serious.

The Minister of Pastoral Production has also requested an urgent discussion of the impact statements issued by the New International Meteorological Organiza-tion (NIMO) Impact Monitoring Unit. The statements indicate that the global cooling may produce world-wide shortages of wool, meat, and dairy products. You will recall that the NIMO Impact Monitoring Unit was established in the old Me-teorological Office building in Wellington in January 1990, so it's an exciting expansion of what the 'old' Meteorological Service was trying to do in the 1970s and 1980s.

Dr Brown. Yes, it certainly is. I remember well how New Zealand pioneered some of the work on the impact of weather and climate on socio-economic activities. Our research obviously made some impressions on the Governing Council of NIMO. The problem in those days was that we did most of the work and others took the credit. However, with the headquarters of NIMO now in Perth, Australia, the north now looks to the south; quite a contrast with the old Geneva days!

B. WEATHER / CLIMATE: INFORMATION OPPORTUNITY

The sensitivity of the international commodity markets to weather and climate information is a clear indication that, in the 'real' economic and business world, weather and climate sensitivity (past, present, and forecast) is a reality. There is also realism in the very difficult areas of disaster relief, and agricultural and energy policies. However, the overall sensitivity - in economic, social, strategic, and political terms - of nations, and sectors of nations,[6] to weather and climate variations, changes, and forecasts[7] needs much more understanding.

The connections between the difficulties of weather and climate forecasting, and the equally complex problems which politicians and social scientists face, point clearly to an even more difficult problem when one tries to link the meteorological and climatological system with the economic, political, and social system. It can of course be done and McKay had some very pertinent comments to make in relation to this question in an editorial[8] in the journal *Climatic Change*:

> Where value can be demonstrated clearly the product will be demanded. Our challenge is to produce practical information that can be readily understood and integrated in a smooth and timely fashion into the planning process. The chances of success in this regard are improved when the planning process is understood - they are much improved when the user is convinced and involved.

These comments emphasize the considerable difficulty facing climatologists and applied meteorologists in convincing decision-makers that there is much more to meteorology and climatology than tomorrow's forecast or the average rainfall in Adelaide, Amsterdam, or Algiers. McKay noted further in the editorial that closer involvement with the user is essential to ensure viability, relevance, and real benefits from new information. Indeed, their interest is, says McKay: 'in more useful information, not in answers to complex problems that they do not understand or complex answers they have to suspect.'

C. THE CHALLENGE AHEAD

As we enter the late 1980's current and projected weather and climatic fluctuations are producing potential, if not actual, significant economic, social, political, and strategic consequences. Our vulnerability to such fluctuations has undoubtedly grown as the world's population has increased and the use of available resources has become more intense. It is now well recognized that particularly in regard to food, energy, and commodity flows, the monitoring and prediction of productivity and consumption at all levels is ultimately associated with the availability, use, and interpretation of the past, present, and future weather.

The foregoing suggests important goals for weather and climate-related activities in the future; many of the goals will involve areas of research and investigation peripheral to the normal activities of many meteorologists and climatologists. The problems are real, and it is believed that meteorologists, climatologists, and geographers, together with experts in other disciplines, can contribute in a very positive and useful way.

For example, there is a need for the development of more rigorous techniques for assessing the *specific* impacts of weather and climate, and in particular determining the degree of impact that climate variability (including the predicted warming from greenhouse gases) has upon all national and several 'regional' economies. This will involve the development of new areas of inter-disciplinary research requiring the skills of a wide range of people from the fields of geography,[9] agriculture,[10] forestry, economics, planning, marketing, political science, and sociology, as well as meteorology and climatology. In addition, the more specialized viewpoints and expertise of people on the 'fringe' of these disciplines must be utilized in order that 'real-world' problems are solved. A good example of this is the thermal mapping and 'open road' package of information produced by a University of Birmingham company.[11]

The multi-disciplinary nature of the climate system also calls for a mechanism to make available much more relevant information to a wide range of people. The need for better climate monitoring systems are also a key aspect of future trends in meteorology and climatology;[12] these and other futuristic aspects of the subject[13] are listed in Table XI.1.

The concerns and implications of the wide variety of items mentioned in this book highlight three key factors: first, climate must be recognized as a resource, and not solely as a factor which imposes *limitations* on agriculture production, settlement patterns, and economic activities; second, there is a need for a much improved understanding of the multitude of inter-relationships between climate and society including health aspects,[14] and in particular in the manner in which changes in one may result in shifts in the other; third, many of the issues of concern have potential if not actual political and strategic implications.

The key factor of the whole weather and climate business discussed in this book is the question of how can the real importance of weather and climate in the market place be achieved. Meteorologists and climatologists have an important role to play in educating and influencing communities and governments of the importance of the atmospheric resource.[15] Nevertheless, the final influencing factor will not be the meteorologists or the climatologists, nor the decision-makers, rather, it is the atmosphere itself which will have the final say[16] if people do not learn to live within their 'climatic income'.

Table XI.1 Climatology - future trends

Observations	Automatic stations Transmission in real-time Analyses in near real-time Immediate availability to users
Computer-to-computer exchange	Climatic historic data Near real-time data
Global/regional monitoring	Enhanced global telecommunications and World Weather Watch system
Real-time exchange	Data Derived products Mapped parameters Commodity-weighted indices
Videotex services	Past information Present data Future indicators
Applications	Area National Regional Global
Research	Continuous monitoring (linked to past) Diagnostic analyses Large-scale modelling Predictions
Impacts	National/regional Advice to governments Political implications Social implications Strategic implications

Source: After Maunder (1984b).

As this book was being compiled a dramatic example of how important the weather/climate 'package' can be even to a highly developed nation occurred in the United States. In this case the severe drought and the associated extreme temperatures which extended over much of the United States and neighbouring Canada in the summer of 1988 brought a virtual avalanche of weather-related reports to the ' front pages' of the media. In some cases these reports were 'linked' to various greenhouse gas/climate change scenarios, and clearly many people within North America came face to face with the reality of a climate which is not always kind. Outside of the United States and Canada other severe weather events occurred which were also noteworthy during 1988, but the severe drought in the

United States in the summer and early fall of 1988 was particularly noteworthily in its political and social impact. This was reflected in the importance which the media placed on the weather events. Typical of these reports were the following:

USA Today   6 July 1988

Soybeans brought USA farmers revenue of $1014 billion dollars last year, twice as much as wheat or cotton. Farmers love higher prices: $9.30 a bushel on the spot market Tuesday (July 5) vs. an average $4.78 a bushel in 1987. But 'higher prices don't help if you don't have a crop,' says Wayne Bennett, president of the American Soybean Association. 'We're facing disaster.'

Financial Post (Toronto) 1 July 1988

Mid-western banks (in the US) are expecting their business to suffer this year because of the long and severe drought, which has slashed the demand for new loans and is holding back the growth of the regional economy. ... Douglas Fisher, vice-president of Hawkeye Bancorp. in Des Moines, Iowa, said spending cuts are already evident in the smaller farm towns as farmers worry the drought could wipe them out. ... 'People have quit buying. ... Car sales are down right now. You can almost measure it to the day the drought started ...'

Columbia Missourian 24 June 1988

With electric utilities feeling the strain as the Mississippi River level continues to fall, the Illinois governor proposed Thursday that the river level be raised by diverting water from Lake Michigan.

Calgary Herald 8July 1988 (Editorial)

Recently announced federal and provincial drought aid plans are still needed. Livestock producers were encouraged by federal Agriculture Minister John Wise's announcement of a $76.5 million drought assistance package. ... In times of drought, government is right to intervene. ... Farm aid can't ever fully substitute for rain, but it is necessary to offset lost income. Otherwise, farmers and ranchers would certainly be facing wreck.

These examples, from the few North American newspapers, show clearly that meteorological and climatological studies - both now *and* in the future - cannot and indeed *must not* be confined to the analysis of atmospheric data. Rather, such studies (whether reported in the media or not) must consider the economic, social, and political aspects of food, health, energy, and well-being, and of all disciplines it is the geographically, environmentally, or economically trained meteorologist or climatologist who can perhaps best cast such studies in a framework that considers not only the earth's surface and the atmosphere, but *also* the economic, social, political, and strategic implications.

To the degree that this is done climatology and meteorology will not only meet the challenge of the 1990's and beyond, but will also emerge as the most influential discipline of the twenty-first century.

Notes

CHAPTER I

1   See Ausubel and Biswas (1980)
2   See Mason (1981)
3   See *New Zealand Geographer*, February 1981
4   See Pippard (1982); also see Mason (1983)
5   For example, the Swedish Meteorological and Hydrological Institute is developing a new Meteorological Information system called PROMIS-(90) , which covers the whole spectrum of predictions from very short range to medium range
6   See Watson (1963)
7   See Maunder (1983)
8   See Goulter (1982)
9   See Baier (1977)
10   See Robertson (1974)
11   See Sewell and MacDonald-McGee (1983)
12   See Meyer-Abich (1980)

CHAPTER II

1   See for example McQuigg and Thompson (1966), Maunder (1968a, 1968b), and Maunder, Johnson, and McQuigg (1971a, 1971b) for five studies in the 'pioneering' 1966-71 period
2   See Maunder (1973)
3   See Ausubel and Biswas (1980)
4   See Mason (1981)
5   Sir John Mason was Director-General of the British Meteorological Office from 1964 to 1983
6   See Maunder (1982a) and also Chapter X
7   As noted by Wallen (1968)
8   See Thornthwaite (1953)
9   See Lamb (1982)
10   See Smith (1961)
11   See McQuigg (1964)
12   See United States Weather Bureau (1964)
13   See Australian Bureau of Meteorology (1965)
14   See Thompson (1967)
15   See Maunder (1965)
16   See Sewell (1966)
17   See Maunder (1970)
18   See Schneider (1976)
19   See for example Glantz (1977a, 1977b)
20   See Phillips and McKay (1980)
21   See Eddy *et al.* (1980)
22   See Center for Environmental Assessment Services (1980, 1981, 1982)
23   See Lamb (1979, 1982) and Wigley *et al.* (1985)
24   See Chen and Parry (1987), and Parry, Carter, and Konijn (1988)
25   See Kates, Ausubel, and Berberian (1985)
26   See White (1982)
27   See Hare (1980)
28   See Epstein (1976)
29   When speaking at the opening of the National (New Zealand) Physics Conference
30   As quoted in the *Bulletin of the American Meteorological Society*, November 1977
31   See Bernard (1976)
32   See Center for Environmental Assessment Services (1981)
33   See Maunder (1986a), Table IV.7, pp.115-18
34   See AIR (Atmospheric Impacts Research) Group (1988)

CHAPTER III

1   See Ashford (1982)
2   See White (1982)
3   See United Nations Environment Programme (UNEP) (1985)
4   See Idso (1984) (a later paper by Idso (1988) suggests however that some viewpoints may be more a matter of intepretation and emphasis than of substance)
5   See Tyndall (1861)
6   See National Research Council (1979)
7   See Tucker (1985)
8   See Monteith (1981)
9   See World Meteorological Organization (1988)
10   See World Meteorological Organization (1988)
11   The Proceedings of this conference are expected to be published early in 1989
12   A report from the Commonwealth Expert Group on Climatic Change and Sea Level is expected to be published late in1989
13   See Clarkson (1988b). Also see *New Scientist* (13 November 1987), *Science* (1 January 1988, and 25 March 1988), and *Scientific American* (January 1988) for recent relevant articles on ozone
14   See Farman, Gardiner, and Shanklin (1985)
15   See Clarkson (1988b)
16   See Clarkson (1988a). Dr Clarkson was head of the New Zealand delegation at the Protocol Meeting in Montreal in September 1987
17   See Fraser (1988)
18   See Lowe, Manning, Betteridge, and Clarkson (1988)
19   Also see footnote 15, Chapter XI
20   See Sewell (1966)
21   See Sewell *et al.* (1968)
22   See Miller (1984)
23   See for example the excellent book on the rhetoric and reality of the subject of Park (1987)
24   See Elsom (1984)
25   See Cess (1985)
26   See Royal Society of New Zealand (1985)
27   See National Defense University (1980)
28   See for example Parry (1985), and Parry and Carter (1983)
29   See United Nations Environment Programme (UNEP) (1987)
30   See Reibsame (1988)
31   See footnote 12
32   See for example Hare (1985a)

CHAPTER IV

1   See Maunder (1970, 1981a)
2   Hallanger (1963)
3   See Mather *et al.* (1981)
4   Bandyopadhyaya (1983)
5   Maunder (1973)
6   See Murphy and Brown (1984)
7   See McQuigg (1972)
8   See Kolb and Rapp (1962), and Lave (1963)
9   See for example Palutikof (1983)
10   See Tolstikov (1968), Mason (1966), and Maunder (1972b)
11   See Mason (1966)
12   See Russo (1966)
13   See Mason (1966)
14   See Wahlibin (1984)
15   See Maunder (1972d)

16 There is also difficulty in obtaining information on the costs of providing meteorological services. For an early review of this problem see Ashford (1969)
17 See Mason (1966)
18 See Curtis and Murphy (1985) for an informative summary of a newspaper survey of forecast terminology understanding

CHAPTER V

1 See Moodie and Catchpole (1976)
2 See Sewell and MacDonald-McGee (1983)
3 See Unninayer (1983)
4 See Sommerville (1987)
5 See Glantz (1977a)

CHAPTER VI

1 See Center for Environmental Assessment Services (1980)
2 See Center for Environmental Assessment Services (1981)
3 See Mason (1966)
4 See Tolstikov (1968)
5 See New Zealand Meteorological Service (1979)
6 See Maunder (1986a), Table IV.3, pp. 96-7
7 See Maunder (1970)
8 See Massachusetts Institute of Technology (1980)
9 See for example Burgos (1979), Fukui (1979), and Oguntoyinbo and Odingo (1979)
10 See Russo (1966)
11 See Linden (1962)
12 See Palutikof (1983)
13 See for example Burton *et al.* (1978)
14 See for example Jodha and Mascarenhas (1985)
15 See Climate Impact Assessment Program (1975)
16 See Eddy *et al.* (1980)
17 See Phillips (1986)
18 See for example Sewell *et al.*(1968), McQuigg (1970), Maunder (1968b)
19 See Eddy *et al.* (1980)
20 See Center for Environmental Assessment Services (1980)

CHAPTER VII

1 See Maunder (1972a) and (1972c)
2 See Maunder (1974)

CHAPTER VIII

1 See Warrick and Bowden (1981)
2 See Anonymous (1966)
3 See Heathcote (1967)
4 See Maunder (1971b)
5 See Perry (1962)
6 See Center for Environmental Assessment Services (1982)
7 See Maunder (1982b)
8 See Winstanley (1985)
9 See Tooze (1984)
10 See Hare (1985b)
11 Examples for dairying, mutton and lamb products, and beef and veal products are given in Maunder (1963)

12   See Walsh (1981)
13   As quoted in New Zealand's *National Business Review* on 10 June 1988
14   For current examples, see the commodity pages of the *Financial Times* or the *Wall Street Journal*
15   See McQuigg Consultants Inc. (1981)

CHAPTER IX

1   See Davies (1960)
2   See for example Stephens (1951), Nye (1975), and Harris (1964)
3   See Johnson, McQuigg and Rothrock (1969)
4   See Maunder (1971c)
5   See Le Comte and Warren (1981)
6   See Maunder (1982b)
7   See Johnson, McQuigg, and Rothrock (1969)
8   See Eberly (1966)
9   See United States Weather Bureau (1964)
10   See Wilson (1966)
11   See Bickert and Browne (1966)
12   See Palutikof (1983)
13   See Steele (1951)
14   See Steele (1951)
15   See Zeisel (1950)
16   See Linden (1959)
17   See Petty (1963)
18   See Maunder (1973)
19   See Gabe (1985)
20   See Maunder (1973)
21   See Linden (1959)
22   See Maunder, Johnson, and McQuigg (1971a)
23   See Maunder, Johnson, and McQuigg (1971b)
24   As stated in the *Bulletin of the American Meteorological Society*, 58: 534 (June 1977)
25   See Flemming (1982)
26   See Troughton (1977)
27   See Taylor (1977)
28   N. Rennie writing in the *Sunday Star* (Auckland New Zealand) on 21 August 1988, stated that in 1960/61 the New Zealand farmer received 54% of the price of a lamb sold on the Smithfield Market in London, whereas in 1975/76 it was only 38%, and in 1987/88 only 19%
29   See Benjamin and Davis (1971)
30   See Russo (1966)
31   See Johnson and McQuigg (1972)

CHAPTER X

1   See Maunder (1977a)
2   For additional information including the regression equations used in the models, see Maunder (1977a), and Maunder (1986b)
3   See Maunder (1978)
4   See Maunder (1980a)
5   See Thomson and Taylor (1975)
6   See for example Maunder (1967) and Thomson (1975)
7   See Maunder (1981b)
8   See Maunder (1982a)
9   See Maunder (1988c)

CHAPTER XI

1   This discussion is based on part of the text of a futuristic discussion between 'Dr Brown' and 'Dr Jones' presented by Dr W.R.D. Sewell and Dr W.J. Maunder, at the Fifth Conference of the Meteorological Society of New Zealand on 11 October 1984 (see Sewell and Maunder, 1985)

2   New Zealand which comprises two main islands (the North Island and the South Island) has currently (1988) one central government

3   Kapiti Island is a delightful bird sanctuary 60 km north of the capital city of Wellington

4   Three items were discussed in the original paper; only one item is discussed here

5   In this connection the comments by Lamb (1988) in his book *Weather, Climate and Human Affairs* on 'The future of the Earth - greenhouse or refrigerator?' are to the point when he says (p. 349) '... Science is not yet in a position to make reliable climatic forecasts over many years ahead, apart from the near-certainty of development of the first stages of the next ice age around 3000-7000 years from now. Our understanding makes it fairly clear that the world-wide warmth of the twentieth century in comparison with the immediately preceding centuries, should be attributed partly to carbon dioxide increase; but it seems that the roughly 50-years-long lull in volcanic activity in the northern hemisphere, which led to a clearer, more transparent atmosphere, also played an important part - it may have been the main part.'

6   See for example the study by de Freitas (1987) on the impact of climate change on New Zealand

7   For an interesting analysis of who subscribes to the *U.S.Monthly and Seasonal Weather Outlook* (published by the NOAA Climate Analysis Center) and what determines the use of the forecasts see Easterling (1986)

8   See McKay (1979)

9   For example, Soons (1987) in an innovative paper notes that '... there are few lines of research in physical geography that to not have some relevance to human activities'

10   See for example Liverman (1987)

11   See Thornes (1987)

12   See Unninayar (1983)

13   See Maunder (1984b)

14   See for example the excellent brochure *Climate and Human Health* published by the World Meteorological Organization (1987)

15   See for example the informative educational booklets on *The Greenhouse Gases* and *Ozone* produced as part of the UNEP/GEMS Environment Library (see United Nations Environment Programme, 1987) or the much earlier papers on weather and climate on economic development in *Finance and Development* (a publication of the International Monetary Fund and the World Bank Group) by Oury (1969) and Kamarck (1973), both of which may have had or will have an influence on decision-makers

16   For example several bridge collapses, including the Tay Bridge (Scotland) in 1879, and the Tacoma Narrows Bridge (USA) in 1940, have been traced directly to a poor understanding of the atmospheric climate

Bibliography

AIR (Atmospheric Impacts Research) Group, 1988: *The Impact of Climatic Variability on United Kingdom Industry*. (M.L. Parry and N.J.Reid, Editors). Report of the Atmospheric Impacts Research Group of the University of Birmingham, Birmingham.

Anonymous, 1966: *Current Affairs Bulletin*. Dept. Adult Education, University of Sydney, 38 (4): 51-64.

Ashford, O.M., 1969: The costs of meteorological services. *WMO Bulletin*, 18: 16-18.

Ashford, O.M., 1982: The launching of GARP. *Weather*, 37: 265-72.

Australian Bureau of Meteorology, 1965: *What is Weather Worth?* Papers presented to the Productivity Conference, Melbourne, Australia, 31 August - 4 September.

Ausubel, J.H., and Biswas, A.K., (Editors), 1980: *Climatic Constraints and Human Activities*. IIASA Proceedings Series, Volume 10, Pergamon Press, Oxford, 205 pp.

Baier, W., 1977: Crop-weather models and their use in yield assessments. *W.M.O. Technical Note*, No. 151.

Bandyopadhyaya, J., 1983: *Climate and World Order: An Inquiry into the Natural Cause of Underdevelopment*. South Asian Publishers, New Delhi.

Benjamin, N.B.H., and Davis, C.C., 1971: *Impact of Weather on Construction Planning*. Paper presented at the National Meeting on Environmental Engineering, American Society of Civil Engineers, St Louis, 21 October 1971.

Bernard, E.A., 1976: Costs and structure of meteorological services with special reference to the problem of developing countries. *W.M.O. Technical Note*, No. 146.

Bickert, C. Von E. and Browne, T.D., 1966: Perception of the weather on manufacturing: A study of five firms. In: Sewell, W.R.D., (Editor), *Human Dimensions of Weather Modification*. University of Chicago, Department of Geography, Research Paper No. 105, pp. 307-22.

Burgos, J.J., 1979: Renewable resources and agriculture in Latin America. In: *Proceedings of the World Climate Conference*, World Meteorological Organization, Geneva, pp. 525-51.

Burton, I., Kates, R.W., and White, G.F., 1978: *The Environment as a Hazard*. Oxford University Press, New York.

Center for Environmental Assessment Services, 1980: *Guide to Environmental Impacts on Society*. U.S. Department of Commerce, Washimgton, D.C.

Center for Environmental Assessment Services, 1981: *Retrospective Evaluation of the 1980 Heatwave and Drought*. U.S. Department of Commerce, Washington, D.C.

Center for Environmental Assessment Services, 1982: *U.S. Economic and Social Impacts of the Record 1976/77 Winter Freeze and Drought*. U.S. Department of Commerce, Washington, D.C.

Cess, R.D., 1985: Nuclear War: Illustrated effects of atmospheric smoke and dust upon solar radiation. *Climatic Change*, 7: 237-51.

Chen, R.S. and Parry, M.L., 1987: *Climate Impacts and Public Policy*. Research Report, International Institute for Applied Systems Analysis, Laxenburg, Austria, 54 pp.

Clarkson,T.S.,1988a: The ozone hole. *World Affairs*, 1st Quarter: 6-7.

Clarkson,T.S., 1988b: The ozone layer, chlorofluorcarbons and the Montreal Protocol. *New Zealand Engineering*, May 1, pp. 7, 9.

Climate Impact Assessment Program (CIAP), 1975: *Economic and Social Measures of Biological and Climatic Change*. Monograph 6., U.S. Department of Transportation, Washington, D.C.

Curtis, J.C. and Murphy, A.H., 1985: Public intepretation and understanding of forecasting terminology: some results of a newspaper survey in Seattle, Washington. *Bulletin American Meteorological Society*, 66: 810-19.

Davies, M., 1960: Grid system operation and the weather. *Weather*, 15: 18-24.

de Freitas, C.R., 1987: Perspectives on the impact of short-term climatic change in New Zealand. *New Zealand Geographer*, 43: 169-76.

Easterling,W.E. 1986: Subscribers to the NOAA Monthly and Seasonal Weather Outlook. *Bulletin American Meteorological Society*, 67: 402 -10.

Eberly, D.L., 1966: Weather modification and the operations of an electric power utility: The Pacific Gas and Electric Company's test program. In: Sewell, W.R.D., (Editor), *Human Dimensions of Weather Modification*. Department of Geography Research Paper 105, University of Chicago, pp. 209-26.

Eddy, A., *et al.*, 1980: *The Economic Impact of Climate*, Vol 1, 1980, (Subsequent Volumes 1-25, 1980, 1981, 1982, 1983, 1984, 1985,1986,1987). Oklahoma Climatological Survey, Norman, Oklahoma.

Elsom, D.M., 1984: Climatic change induced by a large-scale nuclear exchange. *Weather*, 39: 268-71.

Epstein, E.S., 1976: NOAA Policy on industrial meteorology. *Bulletin American Meteorological Society*, 57: 1334-40.

Farman, J.C., Gardiner, B.G., and Shanklin, J.D., 1985: Large losses of total ozone reveal seasonal ClOx/ NOx interaction. *Nature*, 315: 207-10.

Flemming, A.K., 1982: An industry in crisis? *4 Quarter*, 3: 15-77.

Fraser, P.J., (Editor), 1988: *Environmental Health and Economic Implications of the Use of Chlorofluorocarbons as Aerosol Substitutes*. A joint Australian Environment Council / National Health and Medical Research Council document, Australia Government Publishing Service, Canberra, 121 pp.

Fukui, H. 1979: Climatic variability and agriculture in tropical moist regions. In: *Proceedings of the World Climate Conference*, World Meteorological Organization, Geneva, pp.436-74.

Gabe, Masanobu, 1985: Weather information - Valuable economic tool in an era of low growth. *Tokyo Newsletter*, Corporate Communications Office, Mitsubishi Corporation, Tokyo.

Glantz, M.H., 1977a: The value of a long-range weather forecast for the West African Sahel. *Bulletin American Meteorological Society*, 58: 150-8.

Glantz, M.H., 1977b: The social value of a reliable long-range forecast. *Ekistics*, 43: 305-13.

Goulter, J., 1982: Disaster country. In: *Christchurch Star*, 1 September.

Hallanger, N.L., 1963: The business of weather: its potential and uses. *Bulletin American Meteorological Society*, 44: 63-7.

Hare, F.K., 1980: *The Carbon Dioxide Question: Canadian Perspectives*. Unpublished paper prepared for the Climate Planning Board of Canada.

Hare, F.K., 1985a: Climate variability and change. In: Kates, R.W., Ausubel, J.H., and Berberian, M., (Editors), *Climate Impact Assessment: Studies of the Interaction of Climate and Society*. SCOPE, 27, John Wiley, New York, pp. 37-68.

Hare, F.K., 1985b: *Climate Variations, Drought and Desertification*. World Meteorological Organization, Geneva, Publication No. 653, 35 pp.

Harris, D.W., 1964: The relationship between relative humidity, temperature and demand for electric power at peak periods. *New Zealand Electrical Journal*, 37(7): 169.

Heathcote, R.L., 1967: *The Effects of Past Droughts on the National Economy*. Paper presented to A.N.Z.A.A.S. Conference, Melbourne.

Idso, S.B., 1984: The CO_2 climate controversy: an issue of global concern. *New Zealand Geographer*, 40: 110-12.

Idso, S.B., 1988: Me and the modelers: perhaps not so different after all. *Climatic Change*, 12: 93.

Jodha, N.S., and Mascarenhas, A.C., 1985: Adjustment in self-provisioning societies. In: Kates, R.W., Ausubel, J.H., and Berberian, M., (Editors), *Climate Impact Assessment: Studies of the Interaction of Climate and Society*. SCOPE, 27, John Wiley, New York, pp. 437-64.

Johnson, S.R., McQuigg, J.D., and Rothrock, T.P., 1969: Temperature modification and cost of electric power generation. *Journal of Applied Meteorology*, 8: 919-26.

Johnson, S.R. and McQuigg, J.D., 1972: *Application of Linear Probability Models in Using Weather Forecasts to Plan Construction Activities*. Dept. Atmospheric Science, University of Missouri (unpublished paper).

Kamarck, A.M., 1973: Climate and economic development. *Finance and Development* (IMF and World Bank), 10(2): 2-8.

Kates, R.W., Ausubel, J.H., and Berberian, M., (Editors), 1985: *Climate Impact Assessment: Studies of the Interaction of Climate and Society*. SCOPE, 27, John Wiley, New York, pp. 131-54.

Kellogg, W.W. and Schware, R., 1981: *Climate Change and Society; Consequences to Increasing Atmospheric Carbon Dioxide*. Westview Press, Boulder, Colorado, pp. 24-8.

Kolb, L.L. and Rapp, R.R. 1962: The utility of weather forecasts to the raisin industry. *Journal of Applied Meteorology*, 1: 8-12.

Lamb, H.H., 1979 : *Climate: Present, Past, and Future - Volume 2: Climatic History and the Future*. Methuen, London.

Lamb, H.H.,1982: *Climate, History and the Modern World*. Methuen, London.

Lamb, H.H., 1988: *Weather, Climate and Human Affairs*. Routledge, London and New York.

Lave, L.B., 1963: The value of better weather information to the raisin industry. *Econometrica*, 31: 151-64.

Le Comte, D.M. and Warren H.E., 1981: Modelling the impact of summer temperatures on national electricity consumption. *Journal of Applied Meteorology*, 20:1415-19.

Liebhardt, K., 1981: The socioeconomic impact of climate. *Enquiry*, 1(3), Spring. Research at the Unioversity of Delaware, Newark, Delaware.

Liljas, E., 1984: Benefits resulting from tailored very-short-range forecasts in Sweden. *Proceedings Nowcasting II Symposium*, Norrkoping, Sweden.

Linden, F., 1959: Weather in business. *The Conference Board Business Record*, 16: 90-4, 101.

Linden, F., 1962: Merchandising weather. *The Conference Board Business Record*, 19(6): 15-16.

Liverman, D.M., 1987: Forecasting the impact of climate on food systems: model testing and model linkage. *Climatic Change*, 11: 267-85.

Lowe, D.C., Manning, M.R., Betteridge, G.P., and Clarkson, T.S., 1988: Changing atmosphere-changing climate. In: *Climate Change - The New Zealand Response*. Proceedings of a Workshop held in Wellington, 29-30 March1988, Ministry of the Environment, Wellington, pp. 25-30.

McKay, G.A., 1979: Editorial in *Climatic Change*, 2(1),

McQuigg, J.D., 1964: *The Economic Value of Weather Information*. Unpublished Ph.D. Dissertation, University of Missouri, Columbia, Missouri.

McQuigg, J.D., 1970: Some attempt to estimate the economic response of weather information.*WMO Bulletin*, 19: 72-8.

McQuigg, J.D., 1972: *The Use of Meteorological Information in Economic Development*. Prepared for WMO Executive Committee Panel on Economic Development, May, 134 pp. (unpublished paper).

McQuigg Consultants Inc., 1981: The weighted weather index. In: *McQuigg Crop/Weather News*, June 5.

McQuigg, J.D. and Thompson, R.G., 1966: Economic value of improved methods of translating weather information into operational terms. *Monthly Weather Review*, 94: 83-7.

Mason, B.J., 1966: The role of meteorology in the national economy. *Weather*, 21: 382-93.

Mason, B.J., 1981: Review of *Climate Constraints and Human Activities* (Ausubel and Biswas,1980). *Quarterly Journal Royal Meteorological Society*, 107: 743-4.

Mason, B.J., 1983: Predictability in science and society. *Meteorological Magazine*, 112: 361-6.

Massachusetts Institute of Technology (Center for Advanced Engineering Study), 1980: *Climate at Risk*. MTR - 80W 322 - 01, Mitre Corporation, McLean, Virginia.

Mather, J.R., Field, R.T., Kalkstein, L.S., Willmott, C.J. and Maunder, W.J., 1981: Climatology: The impact of the seventies and the challenge for the eighties. *Weather and Climate*, 1: 69-76.

Maunder, W.J., 1963: The climates of the pastoral production areas of the world. *Proceedings of the New Zealand Institute of Agricultural Science*, 9: 25-40.

Maunder, W.J., 1965: *The Effect of Climatic Variations on some Aspects of Agricultural Production in New Zealand, and an Assessment of their Significance in the National Agricultural Income.* Unpublished Ph.D. Thesis, University of Otago.

Maunder, W.J., 1967: Climatic variations and wool production: a New Zealand review. *New Zealand Science Review*, 25(4): 35-9.

Maunder, W.J., 1968a: The effect of significant climatic factors on agricultural producton and incomes: a New Zealand example. *Monthly Weather Review*, 96: 39-46.

Maunder, W.J., 1968b: *An Econoclimate Model for Canada: Problems and Prospects.* Paper presented at the Conference and Workshop on Applied Climatology of the American Meteorological Society, Ashville, North Carolina.

Maunder, W.J., 1970: *The Value of the Weather.* Methuen, London, pp. XXIV - 388, 12 plates, 18 text figures, 56 tables. Also published as University Paperback No. 347.

Maunder, W.J., 1971a: The value and use of weather information. *Transactions of the Electric Supply Authority Engineers' Institute of New Zealand Inc.*, 10-20.

Maunder, W.J., 1971b: The economic consequences of drought: with particular reference to the 1969/ 70 drought in New Zealand. *New Zealand Meteorological Service Technical Note*, No. 192.

Maunder, W.J., 1971c: Temperature forecasts and the assessment of electric power demand in New Zealand. *New Zealand Meteorological Service Technical Note*, No. 195.

Maunder, W.J., 1972a: The formulation and use of weather indices weighted according to the significance of areas: a New Zealand example. *New Zealand Geographer*, 28: 130-50.

Maunder, W.J., 1972b: National economic analyses of responses to weather variations. In: *Proceedings of the Seventh New Zealand Geography Conference*, Hamilton, pp. 207-216.

Maunder, W.J., 1972c: National econoclimatic models: problems and applications. *New Zealand Meteorological Service Technical Note*, No. 208.

Maunder, W.J., 1972d: *Assessing the Value of Weather Information - With Particular Reference to the New Zealand Meteorological Service.* Unpublished paper, New Zealand Meteorological Service, Wellington.

Maunder, W.J., 1973: Weekly weather and economic activities on a national scale: an example using United States retail trade data. *Weather*, 28: 2-18.

Maunder, W.J., 1974: National econoclimatic models: problems and applications. In: Taylor, J.A., (Editor), *Climatic Resources and Economic Activity.* Aberystwyth Symposium XV, David and Charles, London, 1974, pp. 237-57.

Maunder, W.J., 1977a: Weather and climate as factors in forecasting national dairy production. In: *Proceedings of the Symposium on the Management of Dynamic Systems in New Zealand Agriculture*, D.S.I.R., Lower Hutt, pp. 101-26.

Maunder, W.J., 1977b: National economic planning: The value of weather information and the role of the meteorologist. In: *Proceedings of the Ninth New Zealand Geography Conference*, Dunedin, pp. 116-20.

Maunder, W.J., 1977c: Climatic constraints to agricultural production : prediction and planning in the New Zealand setting. *New Zealand Agricultural Science*, 11: 110-19.

Maunder, W.J., 1978: Forecasting pastoral production: the use and value of weather based forecasts and the implications to the transportation industry and the nation. In: *Symposium on Meteorology and Transport*, New Zealand Meteorological Service, Wellington, pp. 107-27.

Maunder, W.J., 1979: New Zealand's real economic climate - evaluation and prospects. In: *Symposium on the Value of Meteorology in Economic Planning*, New Zealand Meteorological Service, Wellington, pp. 203-45.

Maunder, W.J., 1980a: The use of econoclimatic models in New Zealand wool production. In: *Proceedings of the Tenth New Zealand Geography Conference*, Auckland, pp. 22-8.

Maunder, W.J., 1980b: Weather and climatic variations and their significance to man. In: *Proceedings of the Tenth New Zealand Geography Conference*, Auckland, pp. 315-19.

Maunder, W.J., 1981a: The economic climate: fact or fiction? In: *Proceedings of the Eleventh New Zealand Geography Conference*, Wellington, pp. 187-92.

Maunder, W.J.,1981b: *National Econoclimate Studies: A Case Study Using United States Housing Starts.* Unpublished Paper, Center for Environmental Assessment Services, National Oceanic and Atmospheric Administration, Columbia, Missouri.

Maunder,W.J., 1982a: National economic indicators: the importance of the weather. In: Heenan,L.D.B., and Kearsley, G.W., (Editors), *Essays in Honour of Ronald Lister*. University of Otago, Dunedin, pp. 41-60.

Maunder,W.J., 1982b: Canterbury weather patterns: past, present, future. In: Crabb, D., (Editor), *Drought and Drought Strategies*. Rural Development and Extension Centre, Studies in Agricultural Extension No. 1, Lincoln College.

Maunder, W.J., 1983: The weather game. *Weather and Climate*, 3: 2-10.

Maunder, W.J., 1984a: Weather Information and Weather. In: *Proceedings of the 1984 Fertiliser Seminar: Utilizing New Technology*, East Coast Fertiliser Co, Napier, pp. 12-21.

Maunder,W.J., 1984b: Climatology: past, present, future... a personal view. *Weather and Climate*, 4: 2-10.

Maunder,W.J., 1985: Climate and socio-economics. In: *Scientific Lectures Presented at the Ninth World Meteorological Congress, 1983,* W.M.O., Geneva, No. 614, pp. 33-59.

Maunder, W.J., 1986a: Weather impacts and weather managemant. In: Slater, F., (Editor), 1986: *People and Environments.* Collins, London, pp. 441-52.

Maunder,W.J., 1986b: *The Uncertainty Business: Risks and Opportunities in Weather and Climate.* Methuen, London, pp. xxviii-420, 26 text figures, 89 tables. Also published by Methuen Inc., New York, 1987.

Maunder,W.J., 1988a: Global warming - a hot time to come. *Soil and Water* (The Journal of the National Water and Soil Conservation Authority of New Zealand): 24(1): 4-8.

Maunder, W.J., 1988b: Economic aspects of agricultural climatology. In: Monteith,J.L., Elston, J.F., and Mount, L.E., (Editors), *Agricultural Meteorology*. Academic Press, London, (forthcoming).

Maunder, W.J., 1988c: What response choices does New Zealand have? In: *Climate Change - The Zealand Response*. Proceeding of a Workshop held in Wellington, 29-30 March 1988, Ministry of the Environment, Wellington, pp. 155-6.

Maunder, W.J., and Ausubel, J.H., 1985: Identifying climate sensitivity. In: Kates, R.W., Ausubel, J.H., and Berberian, M., (Editors), *Climate Impact Assessment: Studies of the Interaction of Climate and Society*. SCOPE, 27, John Wiley, New York, pp. 85-104.

Maunder, W.J., Johnson, S.R., and McQuigg, J.D., 1971a: A study of the effect of weather on road construction: a simulation model. *Monthly Weather Review*, 99: 939-45.

Maunder, W.J., Johnson, S.R., and McQuigg, J.D., 1971b: The effect of weather on road construction: applications of a simulation model. *Monthly Weather Review*, 99: 946-53.

Meyer-Abich, K.M.,1980: Chalk on the white wall? On the transformation of climatological facts into political facts. In: Ausubel, J.H. and Biswas, A.K., (Editors), *Climatic Constraints and Human Activities*. Pergamon Press, Oxford, pp. 61-74.

Miller, J.M., 1984: Acid rain. *Weatherwise*, 37: 233-9.

Moodie, D., and Catchpole, A.J.W., 1976: Valid climatological data from historical sources by content analysis. *Science*, 193: 51-3.

Monteith, J.L.,1981: Climatic variation and the growth of crops. *Quarterly Journal Royal Meteorological Society*, 107: 749-74.

Murphy, A.H., and Brown, B.G., 1984: Short-range weather forecasts and current weather information: User requirements and economic value. In: *Proceedings Nowcasting II Symposium*, Norrkoping, Sweden.

National Academy of Sciences, 1976: *Climate and Food - Climatic Fluctuation and U.S. Agricultural Production*. A Report of the Committee on Climate and Weather Fluctuations and Agricultural Production, National Academy and Sciences, Washington, D.C.

National Defense University, 1980: *Crop Yields and Climate Change to the Year 2000*. Report on the Second Phase of a Climate Impact Assessment. National Defense University, Fort Lesley J. McNair, Washington, D.C.

National Research Council, 1979: *Carbon Dioxide and Climate: A Scientific Assessment*. National Academy Press, Washington, D.C.

New Zealand Meteorological Service, 1979: *Symposium on the Value of Meteorology in Economic Planning*. New Zealand Meteorological Service, Wellington.

Nye, R.H., 1965: The value of services provided by the Bureau of Meteorology in planning within the State Electricity Commission of Victoria. In: *What is Weather Worth?* Bureau of Meteorology, Melbourne, pp. 87-91.

Oguntoyinbo, J.A., and Odingo, R.S., 1979: Climatic variability and land use: an African perspective. In: *Proceedings of the World Climate Conference*, Publication No. 537, World Meteorological Organization, Geneva, pp. 552-80.

Oury, B.,1969: Weather and economic development. *Finance and Development* (IMF and World Bank), 6(2): 24-9.

Palutikof, J., 1983: The impact of weather and climate on industrial production in Great Britain. *Journal of Climatology*, 3: 65-79.

Park, C.C., 1987: *Acid Rain: Rhetoric and Reality.* Methuen, London, 272 pp.

Parry, M.L., (Editor), 1985: *The Sensitivity of National Ecosystems and Agriculture to Climatic Change.* International Institute for Applied Systems Analysis, Laxenburg, Austria. Reprinted from *Climatic Change*, 1985, 7: 1-152.

Parry, M.L., and Carter, T.R., 1983: *Assessing Impacts of Climatic Change in Marginal Areas: The Search for Appropriate Methodology.* IIASA Working Paper WP-83-77. International Institute for Applied Systems Analysis, Laxenburg, Austria.

Parry, M.L., Carter, T.R., and Konijn, N.T., (Editors), 1988: *The Impact of Climatic Variations on Agriculture.* Kluwer Academic Publishers Group, Dordrecht (in association with IIASA and UNEP), Vol 1., 888 pp., Vol 2., 800 pp.

Perry, R.A., 1962: Notes on the Alice Springs area following rain in early 1962. *Arid Zone Newsletter*, 85-91.

Petty, M.T., 1963: Weather and consumer sales. *Bulletin American Meteorological Society*, 44: 68-71.

Phillips, D.W.,1986: Marketing climatogy for today's user. *Climatogical Bulletin*, 20: 3-15.

Phillips, D.W., and McKay, G.A. (Editors), 1980: *Canadian Climate in Review-1980.* Environment Canada, Atmospheric Environment Service, Ottawa.

Pippard, Sir Brian, 1982: Instability and Chaos - Physical models of everyday life. *Interdisciplinary Science Reviews*, 7: 92-101.

Riebsame, W.E., 1988: *Assessing the Social Implications of Climate Fluctuations.* Department of Geography and Natural Hazards Research and Applications Information Center, University of Colorado, pp. 82. (Written under the auspices of the United Nations Environment Programme as part of the World Climate Impacts Programme.)

Robertson, G.W., 1974: World Weather Watch and wheat. *WMO Bulletin*, 23: 149-54.

Royal Society of New Zealand,1985: The threat of nuclear war: a New Zealand perspective. *Royal Society of New Zealand Miscellaneous Series*, No. 11.

Russo, J.A., 1966: The economic impact of weather on the construction industry of the United States. *Bulletin American Meteorological Society*, 47: 967-71.

Schneider, S.H., (with L.E. Mesirow), 1976: *The Genesis Strategy.* Plenum Press, New York.

Sewell, W.R.D., (Editor), 1966: *Human Dimensions of Weather Modification.* University of Chicago, Department of Geography, Research Paper No. 105, 423 pp.

Sewell, W.R.D., *et al.*, 1968: *Human Dimensions of the Atmosphere.* National Science Foundation, Washington, D.C., 174 pp.

Sewell, W.R.D. and MacDonald-McGee, D., 1983: *Climatic Variability and Change in Canada: Towards an Acceleration of the Social Science Research Effort.* Discussion paper prepared for the Canadian Climate Planning Board, August 1983.

Sewell, W.R.D. and Maunder, W.J., 1985: A day in the life of the weather administrator: 15 March, 1994, *Weather and Climate,* 5: 26-31.

Smith, L.P., 1961: Measuring the effects of climatic changes. *New Scientist,* 12: 608-11.

Somerville, R.C.J., 1987: The predictability of weather and climate. *Climatic Change,* 11: 239-46.

Soons, J.M., 1987: Geography: cultural or physical. *New Zealand Journal of Geography,* December 1987, pp. 8-10.

Steele, A.T., 1951: Weather's effect on the sales of a department store. *Journal of Marketing,* 15: 436-43.

Stephens, F.B., 1951: A method for analysing weather effects on electrical power consumption. *Bulletin American Meteorological Society,* 32: 16-20.

Taylor, N.W., 1977: Short term forecasting in the meat industry. *Management of Dynamic Systems in New Zealand Agriculture.* D.S.I.R. Information Series, 129: 79-87.

Thompson, J.C., 1967: Assessing the economic value of a national meteorological service. *World Weather Watch Planning Report,* No. 17, World Meteorological Organization, Geneva.

Thomson, R.T., 1975: An application of regression analysis to the forecasting of national lambing percentage and per head wool clip. *New Zealand Meat and Wool Boards' Economic Service Paper,* T4.

Thomson, R.T. and Taylor, N.W., 1975: The application of meteorological data to supply forecasting in the meat and wool industry. In: *Symposium on Meteorology and Food Production.* New Zealand Meteorological Service, Wellington, pp. 217-29.

Thornes, J., 1987: Prediction nips icy roads danger in the bud. *Surveyor,* 29 January, pp. 18-19.

Thornthwaite, C.W., 1953: Topoclimatology. In: *Proceeding of the Toronto Meteorological Conference,* T.W. Wormell *et al.,* (Editors), Royal Meteorological Society, London, pp. 227-32.

Tolstikov, B.I., 1968: Benefits of meteorological services in the U.S.S.R. In: 'Economic benefits of meteorology'. *WMO Bulletin,* 17: 181-6.

Tooze, S., 1984: Sahel drought - Call for action. *Nature,* 307: 497.

Troughton, J.H., 1977: Preface to Proceedings of a Symposium on the Management of Dynamic Systems in New Zealand Agriculture. *D.S.I.R. Information Series,* 129: 5-6.

Tucker, G.B., 1985: The global CO_2 problem. *Clean Air,* 19: 5-8.

Tyndall, J., 1861: On the absorption and radiation of heat and gases and vapours, and on the physical connection of radiation, absorption and conduction. *Philosophical Magazine Series,* 4(22): 169-84, 273-85.

United Nations Environment Programme (UNEP), 1985: *Joint UNEP/WMO/ICSU Statement on 1985 Villach Conference* on 'An assessment of the role of carbon dioxide and of other greenhouse gases in climate variations and associated impacts'.

United Nations Environment Programme (UNEP), 1987: *The Greenhouse Gases*. UNEP/GEMS Environment Library No. 1, UNEP, Nairobi (Also No. 2, on *Ozone*).

United States Weather Bureau, 1964: *The National Research Effort on Improved Weather Description and Prediction for Social and Economic Purposes*. Federal Council for Science and Technology, Interdepartmental Committee on Atmospheric Sciences.

Unninayar, S., 1983: *Climate System Monitoring*. Unpublished paper, World Climate Programme Department, World Meteorological Organization, Geneva.

Wahlibin, C., 1954: Use of short-range weather forecasts at construction sites: some results from a recent survey. In: *Proceedings of the Nowcasting II Symposium*, Norrkoping, Sweden.

Wallen, C.C., 1968: Definitions and scales in climatology as applied to agriculture. In: *W.M.O. Regional Training Seminar on Agrometeorology*. University of Wagenigen, Wagenigen, pp. 201-12.

Walsh, M.J., 1981: Farm income production and exports: Part II. In: Deane, R.S., Nicholl, P.W.E., and Walsh, M.J., 1981: *External Economic Structure and Policy - An Analysis of New Zealand's Balance of Payments*. Reserve Bank of New Zealand, Wellington.

Warrick, R.A. and Bowden, M.J., 1981: Changing impacts of droughts in the Great Plains. In: Lawson, M. P., and Baker, M. E., (Editors), *The Great Plains, Perspectives and Prospect*. Center for Great Plains Studies, University of Nebraska, Lincoln, Nebraska.

Warrick, R.A. and Riebsame, W.E., 1981: Societal response to CO_2 opportunities for research. *Climatic Change*, 3: 387-428.

Watson, D.J., 1963: Climate, weather and plant yield. In: Evans, L.T., (Editor), 1963: *Environmental Control of Plant Growth*. Academic Press, New York.

White, R.M., 1982: Science, politics, and international atmospheric and oceanic programs. *Bulletin American Meteorological Society*, 63: 924-33.

Wigley, T.M.I., Huckstep, N.J., Ogilvie, A.E.J., Farmer, G., Mortimer, R., and Ingram, M.J., 1985: Historical climate impact assessments. In: Kates, R.W., Ausubel, J.H., and Berberian, M.,(Editors),1985: *Climate Impact Assessment: Studies of the Interaction of Climate and Society*. SCOPE, 27, John Wiley, New York, pp. 529-64.

Wilson, A., 1966: The impact of climate on industrial growth: a case study. In: Sewell, W.R.D. (Editor), 1966: *Human Dimensions of Weather Modification*. University of Chicago, Dept. of Geography, Research Paper No. 105, pp. 249-60.

Winstanley, D., 1985: Africa in drought - a change of climate? *Weatherwise*, 38: 75-81.

World Meteorological Organization (WMO), 1979: *Proceedings of the World Climate Conference: A Conference of Experts on Climate and Mankind*. Publication No. 5371, World Meteorological Organization, Geneva.

World Meteorological Organization (WMO), 1987: *Climate and Human Health*. Brochure prepared for the World Climate Applications Programme of the World Meteorological Organization.

World Meteorological Organization (WMO), 1988: *Developing Policies for Responding to Climatic Change*. Report prepared by Jill Jaeger *et al*. for the World Climate Impact Studies Programme of WMO/UNEP. WMO/TC - No. 225, World Meteorological Organization, Geneva.

Zeisel, H.,1950: How temperature affects sales. *Printers Ink*, 223: 40-2.

Index